VGM Opportunities Series

OPPORTUNITIES IN
HOTEL AND MOTEL
CAREERS

Shepard Henkin

Foreword by
Darryl Hartley-Leonard
President
Hyatt Hotels Corporation

VGM Career Horizons
a division of *NTC Publishing Group*
Lincolnwood, Illinois USA

Cover Photo Credits:

Front cover: upper left and upper right, Hyatt Hotels Corporation; lower left, Cornell University, School of Hotel Administration; lower right, University of Wisconsin—Stout.

Back cover: upper and lower left, Hyatt Hotels Corporation; upper and lower right, Hilton International.

Library of Congress Cataloging-in-Publication Data

Henkin, Shepard.
   Opportunities in hotel and motel careers / Shepard Henkin.
      p. cm. — (VGM opportunities series)
   ISBN 0-8442-8170-0 (hard) : $12.95. — ISBN 0-8442-8171-9 (soft) :
$9.95
   1. Hotel management—Vocational guidance. 2. Motel management—
Vocational guidance. I. Title. II. Series.
TX911.3.V62H46 1991
647.94'068—dc20   91-9812
                                                              CIP

Published by VGM Career Horizons, a division of NTC Publishing Group.
© 1992 by NTC Publishing Group, 4255 West Touhy Avenue,
Lincolnwood (Chicago), Illinois 60646-1975 U.S.A.
Manufactured in the United States of America.

1 2 3 4 5 6 7 8 9 0 VP 9 8 7 6 5 4 3 2 1

ABOUT THE AUTHOR

Shepard Henkin had a varied and distinguished career in the hotel industry, serving with many highly regarded hotels and chains. His positions included marketing, public relations management, operations, acquisitions, profit-center supervision, and consulting services.

For eleven years, Mr. Henkin was vice-president in charge of marketing with Loews Hotels, a major international hotel chain. He had been president and chief operating officer of Association Services, Inc., a Washington-based hotel consulting firm. In addition, he headed sales and promotional activities for the Governor Clinton Hotel in New York and for the 2,500-room Hotel New Yorker. He also organized hotel and restaurant promotional programs for UMC Industries, a St. Louis, Missouri, conglomerate. Mr. Henkin was vice-president, corporate sales, of Olympic Tower in New York, an unusual condominium complex conceived by Aristotle Onassis. He was also associated with Rockefeller Center, Inc.

Mr. Henkin attended Amherst College in Massachusetts, and was a graduate of the University of Iowa in Iowa City. He also wrote another VGM Career Horizons book, *Opportunities in Public Relations*.

FOREWORD

Each year on Hyatt's corporate anniversary, we close our head-quarters in Chicago, and six hundred of us go into the field to spend the day working as bellmen, front desk clerks, bartenders, house-keepers and reservationists. We come back the next day with sore feet and weary muscles, but above all, we return with renewed respect and admiration for the people working in our hotels—the people who make it all happen.

There's no doubt about it—a career in the hotel and motel industry is hard work, and it often requires long hours. It's not for everyone, but take it from someone who's been in the business for 27 years—a career in this industry is one of the most challenging, most exciting, most rewarding careers you could choose.

Today, perhaps more than ever, a career in the travel and tourism industry holds special appeal. First, the industry is growing. By the year 2000, the travel and tourism industry is expected to be the country's largest employer. Increased opportunities for advancement will exist. Second, the industry is becoming more and more sophis-ticated, with greater demand for qualified individuals with solid management, marketing, and technological skills.

Twenty-seven years ago, I started out as a front desk clerk in a Los Angeles Hyatt hotel. Since then, I've worked in virtually every

v

department at Hyatt, and today I rely on 55,000 hard-working, dedicated employees to help me run the company.

I envy people who are just starting out in the industry. Hotels are like stage sets where the story is rewritten every day, and we always need new performers to add spirit and panache. For those of you accepting the challenge, I wish you good fortune and great success.

> Darryl Hartley-Leonard
> President
> Hyatt Hotels Corporation

CONTENTS

department. The operating management. Top management.

CHAPTER 1

THE SCOPE OF THE FIELD

Hotels and motels have been a part of the American scene from the earliest days of history. From the simple roadside inns of the original colonies, which provided food and rest for weary travelers, to the modern steel and brick giants of today, which are practically cities within cities, hotels and motels have been an integral part of every community.

From individually owned properties the industry has grown in size to a multinational giant. Every year new hotel chains form. This is not only an American phenomenon, but it is common to Great Britain, Ireland, France, Japan, Singapore, Hong Kong, and almost every developed country in the world. Most of the major hotels today are part of international chains. This trend extends to other industries, such as the airlines, real estate firms, and financial organizations which have come into the hospitality field for many reasons, including direction of business, cash flow, ability to increase rates to follow exchange fluctuations, and pure investment.

Hotels and motels are not only places where one can obtain food and a night's lodgings, but also centers of community life, with facilities for meeting, entertainment, communication, and personal services. Their stock in trade has always been hospitality and service, and hotels and motels have made an art of dispensing comfort, pleasing the palate, and creating an atmosphere of home for guests.

Today, across America and worldwide, cities, towns, and villages are dotted with hotels and motels of every kind—from small, simple rooming houses to elaborate fully contained motels, skyscraper hotels, and sprawling resorts providing employment to many thousands. In the United States alone, the hospitality industry is one of the largest of all industries.

The hotel-motel industry is unusual among the major industries of the country in that it is comprised of a great variety of skilled and unskilled occupations. Many of these jobs are common only to the industry; others relate to various outside trades and professions. Those employed in the industry include chefs, managers, plumbers, carpenters, porters, bookkeepers, secretaries, engineers, salespeople, printers, telephone operators, elevator operators, upholsterers, painters, bellhops, accountants, cashiers, waiters, electricians, foreign language interpreters, security people, public relations specialists, and scores of other workers.

We shall concentrate on analyzing the occupations found in the larger hotel and motel operations because, for the most part, these jobs are duplicated in the smaller establishments. Depending on its size and locale, the small hotel or motel performs basically the same functions and services as a larger one, except for having a smaller, less specialized staff.

However, keep in mind that although larger operations offer a greater number of opportunities, the small hostelries offer an excellent training ground for fundamental experience in overall hotel-motel operation. Remember, too, that although many beginning jobs do not require any special educational preparation, a broad education will improve your chances for advancement and give you the ability to perform many necessary duties outside your own sphere of experience.

Dr. Robert A. Beck, former dean of the School of Hotel Administration, Cornell University, and an eminent consultant, describes the challenges and opportunities of the hotel industry.

> The hospitality industry offers today's young men and young women a most interesting and exciting career. Management of a hotel or a restaurant calls for a wide range of capabilities. Guests must be received with cordiality and provided with comfortable, well-designed, and tastefully decorated surroundings. They need appetizing, wholesome food that has been wisely bought, properly stored, skillfully prepared, and graciously served. Various other conveniences in public areas, conference and exhibit rooms, communication systems, and travel services are required for proper guest service. Further, a staff of employees must be recruited, trained, and motivated to provide hospitable service. Moreover, all must be successfully coordinated to return a profit to the establishment's investors. For those wishing a rewarding and challenging life in service to their fellow man, a future in the hospitality field should certainly be considered.

TYPES OF HOTELS

There are many different kinds of hotels and motels. The three major types of hotel operations are commercial, residential, and resort. Commercial or transient hotels make up about three-fourths of the hotels in this country. As there are more than 44,000 hotels and motels, with a total of more than three million rooms, this makes up a sizable number. At $57 billion annual sales combined, they represent a major industry in the United States. These commercial or transient hotels cater to commercial travelers, including businesspeople, salespeople, transient visitors, tourists who spend one or more nights at the hotel. Some of the guests may spend longer periods at the hotel, even though the essential business is still commercial. Commercial or transient hotels that operate public dining rooms and restaurants generally make these facilities available to the general public as well as to the hotel guests. This food business is an important part of many commercial hotel operations.

Another major source of revenue is the convention and meeting business. Newly built hotels are constructed with this in mind and older hotels, when modernized, add public space facilities. Hotels and motels without these meeting and banquet rooms are at a disadvantage competitively.

Residential hotels make up about one-tenth of the total number of hotels in the United States. These hotels provide permanent and semipermanent quarters for their guests. Most of them, though not all, also provide food. Some of them have opened their dining rooms and restaurants to the general public. In general, residential hotels are located in suburban or residential districts. But there are also numerous residential hotels located in or near business sections in order to provide their guests with swift and easy access to and from their businesses.

About one-sixth of the total number of hotels in this country are resort hotels. Resort hotel operation varies greatly depending on size and the hotel's distance from large urban centers. In some resort areas, the hotels are expected to provide only food and lodging, but many large resort hotels could not stay in business unless they also provided sport and meeting facilities. Some of the most famous resort hotels offer magnificent provisions for golf, tennis, swimming, boating, dancing, horseback riding, and planned social activities and entertainment.

Resorts also need to generate business to fill in when regular vacation business tapers off. So today, many top resorts, especially those with huge public space, solicit commercial business in the form of conventions, sales meetings, and incentive tours, especially during out-of-season periods. This is a major source of revenue.

A recent phenomenon is the arrival of conference centers, which are generally located in the suburbs. Fully self-contained, these centers provide state-of-the-art audiovisual and technical equipment and meet all physical requirements for business functions. Located where they are, they can ensure few or no distractions for corporate meetings.

OTHER LODGING

In addition to hotels, there are inns, tourist houses, tourist camps, motels, and rooming houses which also provide lodging, and sometimes food, for guests.

Inns vary greatly in their appearance and type of operation. Some are huge, elaborate establishments which offer all the services provided by hotels; others are small establishments which base their appeal on quaintness, unusual services, or decor. In general, inns should be considered hotels. Their type of operation should be judged, as with hotels, by their size, local customs, and the mood, decor, atmosphere, or period they are planned to convey.

Tourist camps, which include cabins or trailer parks, grew up with the advent of the motor age. As with tourist houses, these camps must locate on or near highways with heavy traffic. But unlike tourist houses, which are generally located in towns and cities, tourist camps are usually found along the highway, outside of city limits. These camps cater to motorists in search of inexpensive lodging. Trailer parks are in themselves a major industry and, like the hotel industry, a growing one. Some tourist camps provide service stations and general stores. Many of the original camps were started by service station operators as sources of extra income. Many tourist camps offer employment opportunities primarily during the summer months, when travel is the heaviest.

The motel was adapted from experience in the tourist camp. A deluxe version of the tourist camp, the motel has become more and more popular with travelers and is becoming an increasingly competitive threat to the hotel industry.

Motels today are as modern and as well equipped as hotels. In many instances, since they are newly constructed, motels are even better than their older hotel competition. Motels provide private baths, radio and television, bellhop service, restaurants, telephone service, valet and laundry service, and will even make reservations for you at your next stopping point. Additional features sometimes make motels more convenient for motorists than hotels. Usually located outside of busy downtown ares, motels relieve the driver of the fatiguing task of trying to park on congested city streets. By allowing motorists to park their cars alongside of their rooms (no longer called cabins), motels allow travelers to save on garage bills and miscellaneous tipping, and they make unpacking and packing every night unnecessary.

Because of their locations along highways, at airports, and even in some downtown locations, motels constitute the greatest competition faced by hotels. These sites are chosen with an eye to highway and air traffic, as well as nearness to newly built industrial sections. The increase in highway and air travel has helped augment the growth of airport and highway motels, each new motel diverting a portion of the business that formerly went almost exclusively to downtown hotels. Motels often have better locations than hotels built in former years and when different traffic patterns existed.

In the early days of the hotel industry, hotels were built largely downtown and quite often near railroad stations. With the decrease of railroad passenger traffic and the move of both industry and offices to the suburbs, these downtown hotels are no longer convenient for the customers they once served.

The move to the suburbs by industry and the subsequent spurt in the building of conveniently located motels have been followed by another trend. The companies patronizing suburban motels have called on the motels to supply public space for meetings and meals. Motels have, therefore, added convention, meeting, and public ballroom space to meet these demands. Here again, motels have become a threat to hotels in this lucrative area. Many hotel organizations consider the sudden advent and popularity of motels so threatening that they have entered the motel field themselves.

Another area of the industry is rooming houses. Rooming houses provide inexpensive lodging for weekly or monthly guests. Most people who choose a rooming house are attracted because of low rents and convenient access to transportation. When rooming houses provide meals for their guests, they are then called boarding houses. These houses do not provide the comforts of a hotel but merely the necessities, including room, linens, bath facilities (generally public), and maid service.

While not major in scope, another important type of hotel operation is the conversion of older hotels into senior citizen residences. Certain downtown hotels that have declined in popularity have been converted into housing for older people, who enjoy the easy access to downtown shopping and conveniences.

Add another recent addition to the industry—the specialized hotel. Conference centers, with their focus on business meetings, are one example. Another is the all-suite hotel, which has proven itself a winner, albeit not a major entry as yet. The all-suite hotel offers suites only, and at the same competitive rates that other hotels charge for regular rooms. This trend toward specialized hotels, individual hotels and chains, should be watched for it, too, should become an important segment of the industry.

HOTELS AND THE COMMUNITY

Because hotels* provide not only lodging and meals but also public rooms and space for meetings, much that is newsworthy takes place in hotels. Depending upon the size of the space available, meetings, conventions, luncheons, social events, charity affairs, and other activities of community and often national interest take place in hotels.

By providing public meeting rooms, hotels perform a valuable service for their communities. Public space in hotels allows many activities of local as well as general importance to take place in communities that would otherwise be unable to accommodate them.

Since the first hotel opened its doors to the public, hotels have been the setting for many of the most important local and national events. Civic and national functions usually take place in hotel ballrooms and famous visitors often stop over at the local hotel. Local celebrities, civic dignitaries, and community leaders can often be found at the hotel, having lunch or dinner, attending social or business functions, or going to civic or service club luncheons and meetings. Many groups hold regular meetings and luncheons at hotels.

If you plan to enter the hotel field, your future will always be exciting and interesting. Whether you work in a small or large hotel, in a small or large city, you will be in the midst of things if you are in the hotel business.

You may wonder why other halls or meeting places have not competed for their share of this business. Hotels, because of their long experience in the hospitality and food industries, can offer service second to none in most communities. In larger cities,

---

*From this point on, we shall refer to all hotels, motels, resorts, and other lodging as *hotels* since the occupational information that follows applies generally to all of these establishments.

restaurants and some halls are providing competition, but none can match the prestige offered by a hotel.

THE OUTLOOK FOR EMPLOYMENT

The future of the hotel industry seems fairly stable and secure for those looking for a career in this field. Factors such as increased business travel and greater foreign and domestic tourism will create demand for more hotel and motel workers. In many areas of the country, there is a great shortage of hotel and motel employees, caused in part by a high turnover rate. This shortage should create good opportunities for those trained in all facets of the hospitality industry. In addition, many thousands of workers will be needed to replace those who transfer to other jobs, retire, or die.

The continuing growth of the entire travel industry will undoubtedly affect all kinds of hotels, meaning increased employment, both temporary and permanent, for all types of workers in the industry. Most of the growth in employment will be a direct result of the need for new workers in the many new hotels and motels that are being built in urban areas all across the country, especially along new highways and in expanding resort areas.

An increase in the number of meetings held by individual companies, industries, and associations has fueled the growth of convention-oriented meeting-space construction at the newest hotels. Meeting space has become a greater source of revenue for hotels than in the past and will influence the growth of the industry in the future.

Air travel also influences the hotel industry. The deregulation of air travel, the concentration of major airlines on destinations in large cities, and the birth of smaller airlines to serve the smaller cities and towns—this all affects the growth patterns of hospitality facilities. Obviously, larger cities and resorts now attract the largest amount of major meeting facility business. Smaller cities have to go after regional and smaller meetings because of the air capacities available.

Because of this increased competition from modern, new hotels, many older hotels feel the need to modernize their facilities. Hotels that are unable to renovate face lower occupancy rates and are often forced to reduce overhead costs, cut back on staff, and curtail services.

From a long-range standpoint, however, the demand for hotel rooms and services is expected to increase as the travel business continues to flourish, and the country's population continues to expand. The greatest rise of employment is anticipated in the motel business, stimulated mainly by the building of new interstate highways and bypasses and increased automobile travel, both for business and pleasure. Personnel with special training will be needed in the front office jobs, as well as unspecialized workers in the back of the house. According to the Bureau of Labor Statistics of the U.S. Department of Labor, there are about 1.6 million people working in the industry, including both full-time and part-time workers.

In a message to readers of this book, the late and well-known hotel industry leader and former president of the American Hotel Association, Frank L. Andrews, stated:

> Regarding the future of the hotel business for the young men and women, naturally I am somewhat biased, having started in the industry as a very young man.
>
> I feel it offers all the advantages any other industry can offer. The success of the industry and of any other industry depends upon the aptitude of the individual, his willingness to work, and perseverance.

INCOME

Since the hotel industry includes workers of almost every occupation, it is difficult to try to estimate the salary one can expect in the hotel industry. Since qualifications for each particular occupation vary, many factors must be taken into consideration. In addition, since many hotel workers depend largely on outside income, such as tips and service charges, the salary scale for their positions does not truly reflect their real earnings.

Salaries also vary according to the local wage scales for the various occupations and the size and location of the hotel. An added feature in estimating compensation is the fact that many hotel jobs carry along with them free meals and sometimes lodging and personal valet and laundry services as well. The latter are true especially of resort hotels where all services such as laundry, valet, meals, and recreation facilities are provided for employees in addition to their rooms.

This book includes salary estimates for each occupation discussed. But remember that earnings vary greatly and these estimates cannot be conclusive. You will find that some hotels provide meals and services for a person employed in a certain category, while another hotel will provide only a salary or wage for someone in the same occupation.

In general, earnings in the hotel industry range from a comparatively small weekly wage (augmented by tips, meals, lodging, and services, depending upon the hotel) to thousands of dollars paid out annually to top executives.

The Bureau of Labor Statistics estimates that average hourly earnings are about $6.84, although wages vary widely based on the type of work and the geographic location of the hotel. You make your own place in the hotel industry according to your own ability and performance.

PERSONAL REQUIREMENTS

Probably the most important personal trait necessary for success in the hotel industry is the ability to get along with all kinds of people under all situations. The people you must deal with in this industry, guests and employees alike, range widely in degrees of education, personal experience, intelligence, business background, nationality, and personal characteristics. You must be able to get along with all of them.

When you take stock of yourself, ask yourself one question. Do you like all people well enough to overlook their idiosyncrasies? If you think you do, then this is the field for you. This does not mean that you must have a "smiley" or "sunny" personality. But it does mean that you must be broadminded, tolerant, understanding, and humane. To paraphrase Kipling, if you can mingle with cabbages and walk with kings, then the hotel business is for you.

Barron Hilton, president and chief executive officer, Hilton Hotels Corporation, in a statement for readers of this text, said:

> I believe the lodging industry offers some of the most personally rewarding careers in American business. For the man or woman seeking an opportunity, our industry offers almost every type of career; from marketing, with its research, sales, advertising, and public relations responsibilities; through service functions in lodging and food and beverage; to the specialist fields of finance, architecture, engineering, and law.
>
> An adequate education is fundamental to one's success in our industry, as it is to one's success in any industry of American business. For those desiring specialized educational training for our industry, many of our nation's largest universities offer outstanding hotel and restaurant management schools. However, I think it well to point out that even such specialized training does not guarantee employment in our industry, but it does highly qualify one to seek such an opportunity. For those having the patience and willingness

to invest a period of employment equivalent to that which they have invested in an education, to learn the practical application of their training, for learning the particular operation of companies they join, and to demonstrate their desire to stand apart in effort and creativity, their success is a foregone conclusion.

My greatest wish is that those entering new careers in our industry find the degree of enjoyment, the sense of accomplishment, and the pleasure of the friendships and associations that I have been privileged to know.

Hotels are essentially service organizations; therefore, the training of hotel personnel focuses on the proper handling of the numerous requests received daily from guests. To a large degree, the public is unaware of the vast training essential to top hotel management in the maintenance of hotel buildings and departments. High schools, colleges, and universities have recognized the hotel industry's needs by instituting various courses designed to give students an insight into what makes for successful hotel operation.

The opportunity for advancement is much greater in this industry than in many others, since the various departments of many hotels are relatively small, and hence sincerity of purpose and efficiency of work are easily recognized.

Any person contemplating a career in the hotel industry should be neat, have a flair for detail, and a willingness to be of service to humanity. This last requirement is not a catch-all phrase; it embodies the ability to listen attentively, have a ready smile, and maintain a reserved manner. Therefore, any person possessing an uncontrollable temper or an inbred shyness must try to overcome these defects if he or she is to make a successful career in the modern hotel. To those feeling qualified to make a career out of the hotel business, the pleasant surroundings, the opportunity to meet new people, and the gratification derived from rendering service are but a few of the rewards of a job well done.

If you are contemplating entering the hotel field, it is best to seek training, provided the training is well organized, and you have the patience and ability to absorb the instruction.

In the past, training programs were unorganized. It was common for someone to be hired as a messenger then promoted to elevator operator, then to bellhop, bell captain, rack clerk, assistant room clerk, room clerk, and so on. This system meant that an employee had to spend a considerable amount of time at each job, absorbing knowledge by watching coworkers. It might be 5–10 years before a new employee was qualified to be a room clerk; after observing the duties of a new job, the employee had to await an opening in the department as well.

Today many training programs are available at vocational high schools and university hotel schools, ranging from one- and two-year programs to full four-year curriculums. These programs will enable you, if you apply yourself, to qualify in a shorter time for a position in the work you like most. The need for cooks, food controllers, bakers, and kindred food workers is great. Hotels today cannot undertake expensive training programs for such jobs because the cost of such programs is prohibitive. Consequently, school is your best choice for initial training in hotel management.

And never feel that you will be training in vain. The vast diversification in the hotel industry, all the way from service to sales, opens the door to so many occupations that you will surely find employment.

Mr. Alan S. Jeffrey, previous director, the Educational Institute of American Hotel and Motel Association, writes that the hotel and motel field offers a future in one of the most exciting industries in the world today.

> If you are seeking an exciting future, enjoy meeting and working with people in a growing industry with good pay, job security, and the opportunity to travel and live in differ-

ent places, you may be just the person who should seek a career in the lodging industry.

There is an excitement about the hospitality business that is like none other. It is interesting, challenging, and rewarding. However, there are times when it is also frustrating. It is fast-moving and hectic. Hotels and motels operate twenty-four hours a day, seven days a week, catering to the needs of people on the move.

Innkeeping, as we sometimes refer to it, is an old and respected profession. As far back as the time of the pilgrims, there were hotels in our country. They were sometimes used to hold town meetings or even religious services. Thousands of years before this, small wayside inns were built along trade routes to provide for the needs of the traveler. A small inn with 100 percent occupancy has been famous for almost 2,000 years.

Today it is not unusual for hotels to contain as many as a thousand rooms or more, though there are many with fewer than twenty-five rooms. Big or small, their purpose is the same—to serve the food and shelter the traveling public. Because of this, ours is considered a service industry.

Providing away-from-home lodging and meals is one of the largest and fastest growing industries in the country today. The need for qualified employees is growing just as fast as the industry. Since more people travel today than ever before, and because of the increasing amount of leisure time most Americans enjoy, hotels and motels continue to be built. This means increased job opportunities.

CHAPTER 2

EDUCATIONAL PREPARATION

No other industry in the world offers its employees so much for so little as does the American hotel industry. No other industry better exemplifies the American way. Here is opportunity, in an industry that can point with great pride to its hosts of self-made leaders—top executives who have risen from humble positions.

Any man or woman who is filled with ambition, energy, and the will to succeed can rise to the highest peaks as a hotel executive and rise there more rapidly than in any other occupation in this country. The annals of hotel history are filled with the success stories of overnight rises to fame. And many of these people learned all they know about the hotel business right in it. Many of the nation's leading hotel executives started at the bottom and worked their way up the ladder of success. They began as assistant waiters, bellhops, room clerks, accountants, and pages. Many leading hotel executives have succeeded without benefit of special training. Years ago, when many of them first started out in the business, few schools or colleges gave courses in hotel management. In those days, hotel employees learned their trade only by apprenticeship or by working for a famous hotel executive and learning his or her system. Today, as hotels have become a major industry, a large number of schools and colleges in the

United States have created special classes or complete courses in hotel work. Educational opportunities range from individual courses to one- and two-year programs to full four-year matriculation.

The complex hotel organizations of today require trained personnel. While many executives in the hotel industry came up the ladder without benefit of special educational or training courses, they grew up with hotels in a period when hotels themselves were growing. Today, although the hotel industry continues to expand and improve itself, it needs properly trained personnel to foster its further maturity.

GENERAL EDUCATION

If you are planning to enter the hotel industry, prepare yourself for the field. Above all, do not neglect your general education. Expand your general studies as much as possible. A good general education will shape you into a well-rounded person and give you the ability to meet people from all walks of life confidently and intelligently.

Include languages in your general studies, especially French and Spanish. Since French is an international language and Spanish is spoken by many foreign business travelers, these two languages are very important in the hotel field. Geography is another good subject to study. Since you are dealing with people who come from all sections of the United States and foreign countries, it is helpful to know your geography. It is good business to know not only your guests, but also the cities and countries from which they come.

If you do not plan to continue your general studies at college, there are many excellent hotel training courses given by high schools and vocational schools. Business schools also offer special courses of study in hotel training. You will find these schools right in your own community with no need (in most instances) to travel daily or live away from home in order to attend. In many instances, you can combine your general high school studies with specialized hotel training. Where hotel training courses are given, consult with your school faculty advisor to see if a combined course is possible.

If you plan to continue your studies in college, complete your general education first, if possible. Here again, you have the choice of combining your general studies with specialized courses of study in hotel management. The individual schools and colleges can best advise you whether such combined courses of study are possible and whether they recommend them in your particular case.

Although it is in your own best interests to complete both high school and college in order to build a good background before undertaking your special hotel training, do not consider this a "must." If circumstances prevent you from completing your education, there are still many opportunities for you to enter the hotel industry and to advance up the ladder while learning the industry from the inside. A large hotel employs a broad cross section of workers in many occupations. It is, therefore, impossible to set up rigid educational requirements for entrance into the hotel industry, since necessary training varies with each particular profession. When you realize that the occupations related to the hotel industry include carpentry, plumbing, electrical work, and other trades, you can understand the variation possible in educational requirements and preparation.

There are many jobs in the hotel industry for which no special education or training is required. These are mostly unskilled and lower paying positions. Hotel management or department heads train many of these employees. These jobs might include those of waiter, maid, clerk, page, housekeeper, porter, or elevator operator.

However, if you would like to be promoted from these positions, it is recommended that you continue your education after hours. In cities where special courses in hotel training are available, it is wise to enroll in these programs. Many men and women have been promoted from these unskilled jobs, and this trend will continue.

If you intend to make a career for yourself in the hotel industry, education and completion of special hotel training courses is almost a necessity. Large hotels and hotel chains give preference to educated employees. They particularly seek employees who have completed special hotel training courses given by recognized schools and colleges. Educated and trained personnel make better hotel employees, and they will become the executives and hotel industry leaders of tomorrow. So, college graduate, high school graduate, vocational school graduate, or plain beginner—continue some form of study or preparation for the future outside of working hours. Success in business must be earned.

Mr. Frank G. Wangeman, for many years senior vice-president of the Hilton Hotels Corporation, and executive vice-president and general manager of the Waldorf-Astoria Hotel of New York, has this to say about education's place in preparation for hotel industry careers:

> If one considers the development of the hotel and inn business, which goes back to the days of ancient empires and practically to the birth of civilization, one comes to the realization that this business of ours has changed more in its complexities in the last 100 years than in all the centuries before. Our tempo of change is destined to further accelerate with the constantly improving modes of transportation.
>
> While the basic concepts of service and graciousness and honor to the guest remain the same as in great periods of culture centuries ago, the way of doing business [has changed] as business in the fashion of yesterday no longer stands up under modern demands; and even what is good enough today will be more than outmoded tomorrow. This,

then, is the challenge of the hotel executive of tomorrow. It offers . . . a great opportunity to come to the fore. The well-trained and aspiring youth will particularly find a calling in the hotel field—for youth, by its nature, is in tune with the times, and our business has to reflect the fashions of the times.

In the memory of many of us, the hotel business has grown from one of the small enterprises to the sixth major industry of the United States. . . . The fact that we have become "big business" is amply demonstrated by looking at the Department of Commerce statistics, and it is in being in "big business" that I foresee the greatest challenges to the rising generation of hotel executives.

The time is already here when employees, even in minor departments, benefit from reading business books developed by the Stanford and Harvard Graduate Schools of Business Administration, thus giving us an indication of the direction in which we grow. Yet, we must never forget that the basic skills in innkeeping will bring us success or failure; however, these basic skills, as essential as they are, will not serve as the future hotel executive's foundation unless they are coupled with modern business methods.

Looking back at the great leaders in our business over the last half century, whether it was Caesar Ritz, E. M. Statler, Lucious Boomer, Conrad Hilton, or others, each and every one was ahead of his time. The future leaders of our business will, of course, also be ahead of their time, which means that they will have to pioneer in fields of scientific and business knowledge that were unheard of in the days of our great predecessors.

I can therefore urge my young friends in the hotel business to equip themselves with the best possible all-around education. This education will bring rewards well beyond their fond expectations; for what is there more thrilling than to be an integral, vibrant part of a great business that encompasses practically each and every phase of human life, and that is bound to grow and further develop with the progress in the various fields of transportation?

PREPARE EARLY

You should contact the school or college or your choice as early as possible to be properly prepared to meet the entrance requirements. You can find out if your preparation is along the proper lines only by contacting the individual schools and colleges and ascertaining their requirements.

Some schools that give classes or complete courses of study in hotel work are limited in the number of students they can admit. This is another reason for your early inquiry.

Write directly to the dean or registrar of those schools or colleges you wish to enter. Ask for detailed information concerning courses of study offered, entrance requirements, registration, tuition fees, and other information. It would be an excellent idea to inform the school in advance of the courses you are now taking or your present educational background. In this manner, you can save time and determine immediately whether you are on the right educational track for your hotel education and training.

See appendix B for a list of schools and colleges that offer studies in hotel management.

APPRENTICESHIP AND TRAINING

Hotels offer greater opportunities for young men and women to apprentice and train themselves than many other industries. In addition few other industries can offer the added convenience of hours that fit in well with school hours. Since most hotels operate on a three-shift system, it is easy for students to work after school hours in apprentice jobs at hotels.

The three common hotel shifts are 7:30 A.M. to 3:30 P.M., 3:30 P.M. to 11:30 P.M., and 11:30 P.M. to 7:30 A.M. In some hotels this timing has been adjusted to the even hours, 8:00A.M., 4:00 P.M., and midnight. The hours from 3:30 P.M. to 11:30 P.M. make good school job hours for student trainees.

On-the-job training is an important part of many courses in hotel work. Many schools and colleges that offer hotel training find no better teacher than a job itself. The opportunity for on-the-job training is open not only to training school students but to all young men and women. Whether or not they are attending special hotel schools, part-time apprentices and trainees are welcomed by most hotels.

Perhaps you are in school and wish to enter the hotel industry without attending a special hotel training school or taking hotel courses right now. Your best bet is to apply for a part-time job at the nearest hotel. Opportunities exist to fill such jobs as bell person, elevator operator, page, key clerk, mail clerk, information clerk, file clerk, office helper, chef's helper, kitchen helper, front office assistant, and waiter. Many students put themselves through high school, college, and hotel training courses by taking part-time or full-time jobs, after school hours, in hotels.

As a matter of fact, a part-time job is an excellent way to discover whether you really like the hotel business. Here is a comparatively easy way to learn about the hotel industry and to decide if you like it well enough to continue your studies in hotel administration.

On-the-job training is highly valued, and in hotel training courses, special credits are given for this work. On-the-job training or apprenticeship can substitute partly for outside studies until such time as you are able to complete a hotel training course.

Many hotels have taken in hand their personnel who started in the field as apprentices with no formal education in hotel work. These employees are attending special training sessions to increase their professional growth. Classes are offered in cooperation with the local Career Development Chapter or directly by the Educational Institute of the American Hotel and Motel Association. This unique educational experience cuts across class lines and helps bring professionalism to individuals who have neither the financial resources nor the time to attend a formal course of instruction. The Educational Institute also offers an individualized home study program that provides persons who desire to advance their career the opportunity to learn at their own pace while still earning at their present position.

Since professional growth never stops, the Educational Institute of the American Hotel and Motel Association offers a certified hotel administrator program. For further information concerning this program, or any of the other fine programs offered by the Educational Institute, simply write to The Educational Institute of AH&MA, P.O. Box 1240, East Lansing, Michigan 48826.

If you think you would like to enter the hotel business; if you feel yourself qualified to enter it; if you are ambitious, energetic, and not afraid of hard work; if you are tolerant, understanding, and like all kinds of people from all walks of life—then let nothing stand in your way.

OPPORTUNITIES

Kenneth J. Hine, executive vice-president and chief executive officer of the American Hotel and Motel Association, has the following to say about career opportunities in the lodging industry.

Because of the many different types of lodging establishments and the many services they provide, there are a multitude of jobs available. The qualifications for these jobs are so varied that men and women with a wide range of educational backgrounds, work experience, and skills can find exciting careers in the innkeeping industry. Further, there are many opportunities for part-time or full-time, day or night, seasonal or year-round, technical or nontechnical positions.

Hotel careers can be divided into these major categories:

(1) *Front Office Staff*—responsible for direct personal contact with the guests, handling reservations, special needs, check-in and check-out.

(2) *Service Staff*—responsible for greeting guests, handling baggage, and assisting with travel plans.

(3) *Accounting*—responsible for tracking financial information critical to the operations of any business.

(4) *Food Service Personnel*—responsible for making every meal a pleasant and enjoyable experience.

(5) *Food Preparation*—responsible for ensuring food is prepared properly.

(6) *Housekeeping*—responsible for maintaining a neat and clean home for visitors.

(7) *Sales Department Staff*—responsible for promotions, handling special arrangements for groups such as meetings, banquets, conventions, and all special events such as weddings.

(8) *Other Departments and Services* including: Security, Safety, Fire Protection, Room Service, Laundry, Dry Cleaning, etc.

A career in the lodging industry offers excellent opportunities for advancement. Lack of experience or education is not a barrier to employment in the lodging industry—it only determines where your career begins. Once you have entered the field, the pace at which you move upward largely depends on your willingness to work hard, the desire to do a good job, your level of enthusiasm and eagerness to advance. On-the-job training programs are plentiful, and excellent correspondence courses are available through the Educational Institute of the American Hotel and Motel Association. Fees for vocational training courses are often reimbursed by your employer.

Because of the size and scope of the lodging industry, there is something for everyone who wants to work in this field. It's a fast-paced growth industry that offers new jobs each year, with excellent job security and opportunities for advancement. Further, you can travel and select where you want to work, the hours, and even the season, if you wish!

Salaries compare favorably with other retail trades, plus there are many extra benefits net reflected in salary. For example, in many cases, at least one meal is furnished, excellent benefit plans are available, and often bonus programs can earn individuals up to 30 percent of their base salary.

PERSONAL ATTRIBUTES

One of the most successful hotel operators in the industry today is Preston R. Tisch, chairman of Lowes Hotels. Tisch, who with his brother, Laurence A. Tisch, chairman of the board and chief executive officer of Loews Corp., has created one of this nation's leading hotel chains, comments on qualities that make for success in the hotel field:

As in any other field of endeavor, anybody contemplating a career in the hotel industry should investigate firsthand the many types of jobs available in the innkeeping field and determine which sort of work he or she is best suited for. A person with a flair for cookery, for instance, would make a poor salesperson, and all the hotel schools in the world would doubtless never make this individual a top salesperson. On the other hand, the proper training, coupled with practical experience on the job, could lead to a well-paying and satisfactory position as a chef.

There are no "easy" jobs in the hotel business. Most of the positions call for long hours and a type of dedication not often found in other lines of work. The best rounded hotel people are the ones who started at the bottom and got a very thorough grounding in all phases of the work from back-of-the-house up. Hotel schools can be a help in certain specialized hotel jobs, but there is no substitute for hard experience.

The good hotel [employee], whether a general manager or bellhop, has to like people to be successful. For after all, it is *people* with whom you will be dealing—not machines or cardboard cartons. I will pay more for the ability to handle people than for any other quality or trait. By people, I mean not only the guests but the other employees in the hotel. Generally, the good host is born with this ability. But, to a certain extent, it can be acquired, and it must be acquired if one is to get ahead in hotels.

Second in qualities necessary to the innkeeping profession I would list attention to detail. Very often I find that the most vehement complaints from patrons are due to seemingly insignificant omissions on the part of staff members. A restaurant guest will wait uncomplainingly in line to get a table at a busy restaurant, but will go completely berserk over a dirty water glass or an overly hard dinner roll. He will accept a smaller room than the one he reserved, but will blow his top because a wash cloth is missing from the bath. The waiter or the housekeeper who is lax in the little things automatically puts the entire hotel in a bad light. Some guests will become so wrought up over minor details that they will never return.

Third, every hotel employee must bear in mind the old axiom that the "customer is always right," even if he is entirely wrong. To attempt to defend yourself against an unjust attack is only natural; nevertheless, you must bear in mind that the complainant has paid good money in your establishment, and, in his own mind, there is nobody more important than he. You can prove he is wrong, but in doing so you are bound to lose him and the friends he might otherwise recommend. The smart hotelier will immediately disarm the guest by agreeing with him and offering to make things right without delay. Of course, there are exceptions to this rule, and those are the ones in which some heavy financial outlay is involved by way of restitution.

Fourth, hotel people who want to make progress in their field should give a little more than the job requires. It is the self-starter, the one who develops new ideas on her or his own initiative, who will amount to something in the long [run]. This is the person we are constantly looking for at Loews Hotels.

CHAPTER 3

HOW TO START OUT

Your next step, after completing your educational program, is to seek employment in the hotel industry.

Once you have completed a course of study in any phase of the hotel business, you will find that, in most instances, the school or college itself will have an employment bureau or will have made arrangements with certain hotels and hotel chains for the placement of graduates. For this reason, registration in some schools and colleges is limited to the number of students the school feels it can place at the end of each school year.

If your school or college does not have any arrangement for placing its graduates with hotels or chains, then you will be on your own. The following procedure applies also to the man or woman seeking employment in the hotel industry without the benefit of formal hotel training.

Most hotels or hotel chains have personnel departments. Write, telephone, or call in person at the office of the employment director (assistant manager, personnel director) of those hotels or chains with which you wish to seek employment. Your goal will be not only to register for employment, but also to get yourself interviewed by the person in charge. If an opening exists, you must "sell" yourself as you would to get any job.

Where there are no openings, request information concerning other hotels or cities where possible openings may exist for you based on your experience, education, or background. Hotel people, especially those in the personnel departments, often know of such openings. If you have made a good impression, chances are that you may receive information concerning other opportunities.

In large hotels, besides contacting the person in charge of employment, communicate also with the heads of those departments for which you might qualify. Departmental heads often hire and fire their own employees. And, in some hotels, not all openings are cleared through the employment office. Some employment offices act as recordkeeping centers and perform routine personnel duties only.

If you can sell your personality and ability to the manager or executive head of the organization, he or she may wish to hire you as a trainee. Many managers are eager to find promising personnel for consideration as future executives. And they have the authority to add to the payroll.

THE AMERICAN HOTEL AND MOTEL ASSOCIATION

There is a hotel association in almost every state. One of the tasks they usually perform for members is to act as a clearinghouse for personnel. They often send out regular lists of available people to member hotels. Communicate with your state associations and with associations in other states. While permanent headquarters for these associations are maintained in some states, in most states the headquarters change each year with the election of new officers. For the correct address of the hotel association in your state, communicate with any hotel in your community.

The American Hotel and Motel Association, located in Washington, D.C., represents practically all leading hotels and motels in the United States. Offering many services to its member hotels—such as legal, accounting, employee relations advice, and other helpful information—is also is a clearing center for specialized requests. If you are in Washington, it might be worth your while to call for information about opportunities in hotels. Or, if you are not in Washington, write to the association. It is located at 1201 New York Avenue, N.W., Suite 600, Washington, D.C. 20005.

Another aid when you are making up a list of hotels to contact is the *Hotel and Motel Red Book,* published each year in June by the American Hotel and Motel Association. This is the bible of the hotels in this hemisphere. The *Red Book* lists hotels in the United States, Canada, Mexico, and other countries. You can use this volume as an address book of job leads.

The *Red Book* list addresses of hotels, describes the local railroad service, and provides detailed information about each hotel: the number of rooms, whether the hotel is a summer, winter, or all-year-round operation, whether the plan of operation is American or European, the minimum room rates, and the names of the managers. In addition, the *Red Book* provides the names of the officers and directors of the American Hotel and Motel Association, their affiliations and addresses, and a list of affiliated industries.

You do not need to purchase the book to use it. It can be found at most libraries and at the registration desk of most hotels. Most hotels will be glad to permit you to look at their copy and make notes.

JOB INTERVIEWS

There are two words that should guide you not only when you apply for your first job in the hotel industry, but through your entire hotel career. Those two words are *common sense*. If you use common sense, you will do well at the start of your career and later on.

When applying for a position in a hotel, remember that hotel work is service work. A hotel's business reputation depends upon the quality of service it offers its guests. Service is best performed by people who are clean and neat. In a hotel, the personnel must always be polite, speak correctly, and use good manners.

Keep these points in mind when you apply for a position in a hotel. If you realize the interviewer's priorities, your own common sense should tell you how to act. You are being judged on your intelligence, your appearance, your manners, and your willingness to learn. Do not let "hot-headed," impulsive emotions rule you. The interviewer is looking for a level-headed, self-controlled, flexible person. He or she is looking for someone who can adapt to changing situations and get along with different kinds of people.

Above all, when you apply for a position in the hotel field, remember your appearance. One of the most important requirements in the hotel business is good appearance. Hotel people, by the very nature of their work, are required to be well groomed at all times. The hotel industry can probably claim the best-dressed people of any career field. You cannot expect to make a good impression when interviewing for a hotel job unless you are neat, clean, and appropriately dressed. Good grooming makes sense.

PROMOTIONS

As previously stated, the history of the hotel industry shows that the path to success lies wide open for ambitious, intelligent, energetic people. Many of today's top hotel executives have come up from the ranks, some starting as far down that ladder as assistant waiters, bellhops, and clerks. Tomorrow's hotel leaders may be an assistant waiter in San Francisco, a room clerk in Dallas, an accountant in Philadelphia. Even if they never become top executives, beginners are often promoted to more responsible positions. Housekeepers often start as maids, chefs as apprentices, restaurant managers as assistant waiters. The opportunities are there. It is up to you to take them.

The length of time between promotions in the hotel industry varies. There is no set schedule or plan of advancement in most hotels. The only organizations where regular promotions are given are those conducting executive-training or exchange programs. In the former, someone being groomed for executive work will be rotated into different hotel departments to become familiar with the operations of the hotel. In an exchange program, hotels exchange department or subdepartment heads with one another in order to share ideas and learn from each other's operations.

In general, there is greater turnover in a large hotel than in a small hotel. Accordingly, swifter advancement is possible in the large hotel because openings occur more often, and changes are made to fill vacancies. Mathematically, the law of averages (deaths, retirements, resignations, and transfers) will operate more to your advantage in an organization with many employees.

Management in most small hotels is identical with ownership. This limits your future prospects unless you can raise enough capital to buy or become a partner in a hotel. Most large hotels, by their size alone, represent huge investments. Very few people in the United States have sufficient capital to purchase or build a large hotel. Most large hotels are, therefore, owned by corporations representing huge financial investments of banks, insurance companies, or joint stock companies. Some hotels have been financed by public stock issues. Since the large hotels are generally controlled or managed but seldom owned completely, their corporate structure creates opportunities that would not exist in a privately owned enterprise. Corporations offer greater opportunities for advancement and often make top posts available to rank outsiders.

Uppermost positions in the hotel industry are attained only after many years of managerial and executive experience in the industry. The larger the hotel, the more experience will be required. There is actually a great difference in managing a medium-sized hotel and operating a huge edifice of a thousand or more rooms with many public and dining halls. You can get to one of the top hotels posts only after you have had considerable experience in larger hotels.

Advancement in the hotel industry is unique and quite peculiar. Comparatively rapid, it does not follow a regular pattern, and it may be indirect. In most industries, employees advance or receive increases only after they have spent long periods of time in each position they hold. And in private industry, advancement is more commonly indicated by salary increases rather than a change in position. A driller in the oil industry keeps receiving pay increases, but no one would think of promoting the driller to assistant credit manager. A post office delivery person receives automatic pay increases, but no postmaster would promote the delivery person to a higher position as engraver in the State Department.

In the hotel industry, employees who merit advancement step into positions higher up in rank and salary. But this step up may lead the employee into an entirely different department. The advancement may even mean a move to another hotel, sometimes in a different city or country.

CASE STUDY

Let us consider the hypothetical case of Jonathan Doe. Jonathan's first job in the hotel industry was as a bellhop at the

Metropolis Hotel. His initial salary was $200 per week. After six months, the bell captain noted on Jonathan's record the fact that Jonathan was efficient at his job, clean, neat, intelligent, and willing to learn. Two weeks later when the assistant bell captain left for a position at another hotel, Jonathan was promoted. His salary now became $250 a week, and he had a title.

Some time later, the chief of service needed to replace his assistant, who had been appointed room clerk. The next three employees in line for the job declined, and the post was offered to Jonathan. Jonathan Doe was now assistant chief of service at $325 per week.

Note that each time Jonathan received a wage increase, he also moved up the ladder. His next promotion could be to room clerk or assistant credit manager, neither of which follow his experience thus far in the hotel. In the course of events, Jonathan might be offered a position at another hotel. Such transfers are part and parcel of the hotel business. This promotion process is repeated in almost every department of the hotel.

There are many opportunities for advancement, but Jonathan, and the thousands like him, do not sit and wait for opportunity to arise. They take outside training courses. They contact department heads in the hotel, asking to be considered for openings. Hotel policy usually gives preference on job openings to present employees who keep working efficiently and knocking on the doors to success.

CHAPTER 4

THE FRONT OF THE HOUSE

In hotel jargon, ''the front of the house'' refers to those departments that deal directly with, and are seen by, hotel guests. The front of the house departments also include management departments. Service, front office, accounting, credit, office management, security, personnel, banquet, advertising, public relations, sales, resident management, executive management, and all subdivisions of these departments make up the front of the house. While there are important positions in the back of the house, most top positions and all executive posts lie in the front.

Whether you start your hotel career in the front or back of the house depends on your likes and dislikes, your educational and training background, and your ability and skill in the various professions. But regardless of which side you choose to enter, the opportunities are equal. Only the paths along which you rise will be different.

Even if you start in the front and proceed up the ladder, you will find that knowledge of back-of-the-house operations is most important. Somewhere along your advance in the hotel business, you will need to study, if not actually practice, back-of-the-house duties and operations. Study courses in food operations, housekeeping, purchasing, and other back-of-the-house departments.

Try to work in the various departments, if possible. But round out your hotel experience and training with an all-around background in both front- and back-of-the-house operations.

SERVICE DEPARTMENT

The service departments of most hotels offer beginners in the hotel industry wonderful opportunities for starting their careers. In addition, they are an excellent means for advancement because service department jobs often are stepping-stones to the top of the ladder.

A hotel's entire function is to provide service. However, certain duties that deal with personal service provided for guests entering or leaving the hotel are grouped together and performed by a separate department set up for that purpose. This department is called the service department—sometimes known as the concierge. It includes bellhops, elevator operators, door attendants, and others.

The service department is headed by the superintendent of service. In some hotels, this job is called assistant manager in charge of service. In other hotels, the position is titled chief of service. Under the superintendent of service are such employees as bellhops, door attendants, elevator operators, porters, and checking room attendants. In some hotels, duties of baggage and washroom attendants and pages are supervised by the superintendent of service.

A recent trend has been the rise of the office of the concierge. A European custom for many years, this has emigrated to the United States and is fairly common in most major and luxury hotels. The concierge and superintendent of service positions may be the same in some instances, but mostly the concierge is a sole entity, similar in many ways to the old hospitality desk that once existed in most hotel lobbies. The concierge will arrange special requests, whatever they may be; and in hotels dealing with foreign visitors, will speak more than one language.

Superintendent of Service

The members of the service department provide guests with their first impressions of a hotel. The treatment given guests by the door attendant, the bell person who takes their bags, and the elevator operator influences guests' opinions of the hotel. The responsibilities of the superintendent of service are therefore great, but the job also offers fringe benefits (such as free meals) and good opportunity for advancement.

The superintendent of service is responsible for hiring, instructing, disciplining, and discharging employees in the department. The efficiency of service employees will depend to a large extent upon the efficiency of this person's instruction methods, his or her own personal hotel experience and background, and the type of personnel he or she employs.

The superintendent must ensure that everyone in the department does a good job. Door attendants must be prompt in opening doors of automobiles, ready to help guests in and out of automobiles and cabs, and willing to carry baggage from curb to door where bellhops will pick it up.

Elevators must be operated safely and on the best possible schedule. The operators must be neat and clean, must call off the floors promptly, and must be polite in their dealings with guests. Self-service elevators must be watched and regulated.

It is the duty of package room attendants to ensure the safe delivery of packages to guests.

Bell persons must be alert and intelligent and must respond quickly to the wishes of guests. They must be well trained in hotel procedure, such as hanging clothes in closets, opening windows, checking bathroom supplies and facilities, and checking rooms for completion of proper maid service.

Most superintendents of service have risen to their posts after years of experience. A recent survey shows that the average time required to reach this office is about ten years. Occasionally, front office clerks are promoted to this position. You can also rise to this office from the job of bell captain, head baggage porter, or other jobs in the hotel. However, the office of superintendent of service is not a last stop. It is a stepping-stone job, and many superintendents have advanced to higher positions in the same or other hotels.

Bell Captains

Found in most medium-sized and all large hotels, the position of bell captain is the second ranking job in the service department. After superintendent of service, it is the job most sought after in this department. And in some of the larger hotels, people would prefer this post to that of superintendent because of its financial and other advantages.

Bell captains attain their posts only after years of experience. Most positions here are filled by promoting a bell person. Definitely a stepping-stone position in the hotel organization, the position offers opportunities for operational experience.

It is the duty of the bell captain to keep time records of all bell persons, to instruct all new employees, to arrange the immediate dispatch of bell persons on guest calls, to rate the bell persons fairly so that all share evenly in the tips, and to assign bell persons efficiently so that all incoming guests are met and all guests' requests are complied with. The bell captain is also responsible for interviewing new job applicants, investigating and adjusting guests' complaints relating to the work of the department, and deciding whether unusual guest requests should be filled. An efficient bell captain can make the difference between good and bad service for the hotel guests.

The bell captain's staff includes bell persons and sometimes pages. Bell persons perform a multitude of tasks. They are charged with ushering incoming guests to their rooms and carrying their baggage. Guests will ask them for local and other information which they must be prepared to furnish. Bell persons sometimes also help set up rooms and bring ice water, food, or other items requested by guests in hotels where these are not furnished by the hotel itself. Bell persons are occasionally called upon to perform special duties like delivering letters or packages (with the special permission of the bell captain), picking up theater tickets, making travel arrangements, and many other personal service chores.

Bell persons must be born diplomats. They must be able to judge people and know how to make them feel comfortable and at home in their hotel room. As the first connecting link between the guest and the hotel, the impression they create is important.

Positions as bell persons are secured by making application to the personnel department or bell captain. Some hotels select their bell persons from the ranks of elevator operators or starters, while others employ bell persons who have had experience elsewhere. But the methods of employment differ and depend upon the policy of the hotel in question. Some hotels employ bell persons who have had no experience at all. In communities where courses in bell person training are given by schools, hotels, or hotel associations, local hotels may require their bell persons to attend the courses either before or during employment.

There are many opportunities for bell persons to advance. The next step is promotion to bell captain, then superintendent of service, and then up to the various managerial posts. Bell persons may sometimes move to other hotels where better opportunities exist.

Very good opportunities for entering the ranks of bell persons exist in resort hotels where virtually new crews are hired every season or where large turnovers are common. After gaining experience there, one can transfer to a commercial hotel. Many bell persons start as elevator operators or housekeepers and work their way up. The length of time it takes to advance depends upon the size of the hotel's staff and the rate of turnover.

Bell persons usually work eight hours a day, six days a week. In large hotels, the three-shift system is employed. At the start, therefore, you may find that the beginners are given the night shift, since hotels provide service 24 hours a day.

Head Baggage Porters

In larger hotels, in addition to bell captain, there will also be a head baggage porter. Although we have previously listed many duties of the baggage porter as those of the bell captain and bellhops, this work is handled exclusively by baggage porters in larger hotels. Setting up rooms, supplying travel information, buying transportation tickets, arranging shipment of express articles, and handling baggage and suitcases of guests who are departing are the exclusive duties of the baggage porter in the larger hotels.

The head baggage porter must keep time records of all employees in the department; interview, instruct, discipline, and discharge employees in the department; rotate the staff on calls equitably; and in general perform the same supervisory function as the bell captain.

One of the head baggage porter's primary responsibilities concerns making transportation arrangements, shipping express articles, and buying transportation tickets. For this reason, this person is also commonly referred to as the transportation clerk. His or her office may be called the transportation desk.

Baggage porters are generally appointed from the elevator operating staff, housekeeper's ranks, or other departments of the hotel. Occasionally, a hotel will appoint a baggage porter to this job who has had no outside experience. People with experience at other hotels are also considered.

Baggage porters aspire to be head baggage porter. Most porters reach this position only after several years experience. The next step up the ladder is appointment as superintendent of service. However, this promotion more often goes to the bell captain than to the head baggage porter. From superintendent of service, the path leads to front office or managerial positions.

The working hours and conditions of employment for baggage porters are usually the same as those for bellhops.

Other Service Department Functions

Service departments of hotels will differ, depending upon the size of the hotel and staff and the operation policy of the hotel's management. Departmental setups will vary from hotel to hotel.

Accordingly, some hotels offer additional opportunities for employment in the service department. These positions include door attendant, checking attendant, porter, page, secretary to superintendent of service, lobby attendant, rest room attendants, shoeshine attendants, and others. For further information, apply to the personnel department or superintendent of service for details of other jobs.

With the exception of the head of this department, hours of work in the service departments of most hotels are based on the three-shift system. Hours will vary, depending on the size of the hotel and nature of its operation, and this three-shift system might not then apply. In general, however, employees in this department work about eight hours a day, five or six days a week.

Pay also varies, depending upon the size of the hotel and the city. Bellhops and baggage handlers' earnings are roughly between $4.00 and $5.50 per hour plus tips in smaller cities. Bellhops and porters make more money in larger hotels or resort hotels where more services are demanded. In these settings, their earnings may run as high as $500 weekly and more, including tips.

The average wages received by superintendents of service can range from $300 to $400 weekly in smaller hotels. More money is earned in larger hotels. Bell captains' and head baggage porters' incomes average about the same. Since transportation services are often provided by the head baggage porter, this person's income can sometimes be greater by virtue of large tips received on these transactions.

Because most, though not all, earnings in this department are augmented by tips and side money, one cannot consider the average wage as the complete remuneration. In general, earnings of service department employees run much higher than their base average wage scales would indicate. Earnings here may vary from city to city and hotel to hotel. As indication of the potential earning capacity in these service jobs, people who have been bell persons or baggage porters for many years in some of the larger hotels have refused promotions many times, preferring their current positions.

THE FRONT OFFICE

The entire responsibility for processing reservations, registering guests, and keeping records of room vacancies is in the hands of the hotel's front office. It must conduct these functions efficiently so that the front office manager always has enough information to make firm reservations for guests without overbooking hotel facilities. In addition, the front office performs all tasks related to registering and keeping track of guests, including providing keys and mail service.

One of the most important positions in the hotel is the front office manager. This person is charged with the responsibility of estimating the volume of future reservations, preparing for busy seasons, organizing all departmental functions so that they operate efficiently, and maintaining a close check at all times on occupied and available rooms, and firm reservations. The front office manager must keep all departments constantly in check with one another and well balanced.

Promotion to this position is generally made from front office clerks, assistant managers on the floor, credit office personnel, or other workers. Occasionally hotels will hire front office managers from outside the hotel. From front office manager, the next step up is toward a management executive post.

Under the front office manager are room, rack, and reservation clerks; key, mail, and information clerks; floor clerks (also known as assistant managers on the floor); hospitality department workers and secretaries; filing clerks, typists, and other clerical workers.

Front office managers, because of the importance of their work and the large number of employees they supervise, must have much hotel experience, ability, and mature judgment. Their authority over rooms is second only to that of the director of sales and the manager.

Front Office Clerks

A front office clerk generally has at least a high school education and has completed some courses in hotel training either before or during employment. Although many positions in the front office do not require higher education or special preliminary training, the opportunities offered in the department induce many front office employees to enhance their job experience with hotel training courses. This is especially true with front office clerks as their job is considered a stepping-stone for managerial positions.

Front office clerks perform various duties. In smaller hotels, the front office clerk, or manager, may perform all front office duties. In larger hotels, the work is departmentalized. The front office clerks consist of room clerks who sell rooms and follow through on all functions of guest registration; rack clerks, who enter the names of newly arrived guests on the rack and clean out names of departures; reservation clerks, who acknowledge and make reservations by phone, letter, fax, or telex; and various other subdivisions, depending on the size of the hotel and its staff. In many hotels, data processing has eliminated the need for a rack clerk. An office clerk does all check-in and check-out duties using software programs on a computer.

In general, front office duties include the mechanical processing of reservations, sale and registration of rooms, keeping room racks accurate and up to the minute, furnishing guest keys, and handling complaints about rooms or other accommodations. Front office clerks also receive and forward mail, give information about guests registered or expected (where permitted), and provide local information concerning room rates and times of departure.

While front office clerks are sometimes employed directly from outside applicants, it is general hotel practice to fill these openings with other staff employees, such as bell persons, credit workers, clerical employees, or other personnel. For the more responsible jobs in this department, people with similar experience at another hotel are often hired.

Since you may be promoted from front office clerk to manager, at least a high school education, and preferably a college degree, is recommended for success. If you cannot continue your general education, you should definitely complete special courses in hotel training given by the schools and colleges listed at the back of this book. You can complete special hotel training courses while you are employed, if necessary. Correspondence courses are available as well.

The three-shift system usually prevails in the front office. The shift employing the fewest workers is the night shift, since most new guests arrive during the daylight hours, or before midnight at the latest.

In small hotels, the owner-manager may handle the duties of the front office with or without an assistant. There is more opportunity for advancement and for obtaining knowledge in a large hotel than in a small one. The larger hotel, because of its size and scope of operations, offers many more chances to perform hotel duties.

Front Office Assistant Managers

In larger hotels, front office staffs include assistant managers. They are on duty "on the floor" (the lobby floor). Assistant managers are troubleshooters and ambassadors of good will for the improvement of guest relations.

Their duties are managerial in scope. Representing the management, they handle complaints from guests and assist in straightening out problems in any emergencies that may occur. Regular duties of assistant managers include helping guests make reservations at hotels in other cities, changing guests' rooms as requested, notifying the security department of disorderly or undesirable characters spotted throughout the hotel, assisting with guest registration at rush check-in hours, and helping to register special guests quickly or quietly when requested. Assistant managers see that operations at the front office, in the lobby, and throughout the hotel are functioning properly and that the guests are satisfied.

Although assistant managers are authorized to assume managerial status in emergencies and other situations, they are responsible to the front office manager, and the major part of their duties concern front office operations. Assistant managers on the floor serve a useful purpose, for in addition to expediting guest arrivals and registrations, they also relieve the manager and executive assistants of the many minor problems that occur daily in hotel operation.

Members of the front office staff are usually appointed assistant manager. Front office clerks, chief room clerks, and others are next in line for the position.

Assistant managers work the same hours as the rest of the front office. Since they are a step higher on the ladder of front office operations, the educational and training requirements are the same or greater than those for front office clerks.

Mail and Information

The mail and information department offers excellent opportunities for beginners in the hotel industry. Although the duties of this department are assumed by front office clerks in smaller hotels, this work is more specialized in larger hotels. The duties of this department include handling incoming and outgoing guest mail, supplying information concerning room numbers of guests, clearing such room information for the telephone department and other hotel departments, maintaining guest room key racks, and furnishing guests with room keys.

Mail and information is an excellent department in which to begin a hotel career. Many young men and women are employed in this phase of work, and the department offers good opportunities for advancement. Educational requirements are essentially the same as those for front office clerks. The work here is not necessarily skilled, and what training is required is usually given by the head of the section. However, here the better your education and training, the better background you will have for advancement.

As hours here are based on the same three-shift system used by the rest of front office employees, openings are possible for after-school work. Applicants with no previous experience can find jobs. Occasionally, personnel from other departments are employed at this desk when they are being considered for further promotions.

Hospitality Department/Concierge

The larger commercial and resort hotels have hospitality departments. In smaller hotels, the duties of this department are performed by room clerks, assistant managers, bellhops, or other employees who come in contact with guests.

Sometimes the function of hospitality personnel is to act as hosts for guests of the hotel. The duties include providing guests with information about local points of interest; keeping daily listings of local motion picture and theatrical entertainment; providing special services, such as baby-sitters, companions, and personal maid service; providing radio and television tickets for guests in cities where such programs emanate; arranging sightseeing tours; and helping guests make reservations at hotels in other cities.

The hospitality staff is usually supervised by an assistant manager (front office) on duty. The usual hours of work are during the two daytime shifts of the three-shift system. As in most hotel jobs, opportunities for promotion exist; work in the hospitality department can lead to sales, front desk, and other jobs.

For work here, one should be well informed about local points of interest, have good knowledge of nearby highways, keep abreast of play and motion picture reviews, and in general be in touch with all local social, church, theatrical, and other such events.

Hours and Earnings

Most front office employees are on the three-shift system. While this system of changing hours from one week to another may be irksome at first, most hotel employees become accustomed to it.

From the front office there is often greater opportunity for promotion than from any other department of the hotel. The hotel business is primarily one of selling rooms, food, and liquor. Here in the front office, you are face-to-face with guests, their problems, their complaints, and their likes and dislikes. You can watch the hotel as its rooms empty and fill on charts, racks, and computer screens before you. Many of today's hotel executives started out in front office positions.

Average front office earnings vary, depending upon the size of the hotel, the number of employees, and the size of the city. Typical salaries for front office clerks range from $4.50 to $9.25 per hour for an eight-hour day and a five- to six-day week. Earnings of apprentices, mail and information, and key and other clerks may start lower and vary, depending upon locations. The salary of the front office manager is well above the departmental average, but earnings vary greatly in this department. In some hotels, certain front office jobs include meals as part of the remuneration.

THE ACCOUNTING DEPARTMENT

Although its members are not generally in direct contact with guests, the accounting department is included under front-of-the-house operations because much of the work is managerial and accounting executives often advance to top hotel positions. The accounting department is frequently a separate department reporting directly to the manager.

As in other businesses, the accounting department supervises the financial affairs of the organization. Accounting duties include fiscal policy and planning, maintenance of fiscal records and accounts, preparation of regular periodical and annual financial statements, control of expenditures, and the recording of income, maintenance of bank accounts, and handling of payrolls.

In small hotels the owner-manager may keep a set of simple books, regularly checked or supervised by outside accountants. In larger hotels, however, accounting operations are huge and complicated and require large staffs to maintain them.

For work in the accounting department, you will need special education and training. At least a high school education is required, and if you plan to further yourself in this specialized field, it will be necessary to complete accounting studies and perhaps even become a certified public accountant. Top spots here, as auditor or controller, require accounting backgrounds.

Howard P. James, former chairperson and chief executive officer of the Sheraton Corporation, writing for readers of this book, says:

> A job in the hotel business can indeed take many directions!
>
> In the last few years, hotels have modernized their procedures dramatically and reservations are now handled instantly and make use of computer technology as do modern payroll systems, and even the housekeeping and engineering system in larger hotels.
>
> The hotel business offers openings and career potentials for a wide variety of talents. It is also a business which offers much opportunity for development on the job.

The larger hotels and hotel systems encourage employees to develop their skills and abilities. Many of them provide some schooling themselves or provide funds to assist employees in further training.

The American Hotel and Motel Association offers correspondence courses in every aspect of hotel work. Many accredited colleges and universities have courses leading to a degree in hotel management and also give brief summer courses which are open to working hotel employees. There are also private schools which offer courses.

The hotel business is demanding in its hours of work, some of its busiest periods being when the majority of working people are at their leisure, such as evenings and holidays. Sometimes a part of the opening ceremonies of a new hotel is throwing away the key to the door, symbolizing the fact that the hotel will henceforth be open to the public 24 hours a day.

Historically an ancient trade, the business of innkeeping becomes daily more modern in its techniques. However, because it is essentially a "people business," it maintains the fascination that will always attach to any enterprise having constantly changing personal relationships.

Controller

This is a highly specialized position, and most accountants reach it only after many years of experience, preferably in the hotel business. The controller—or chief accountant, as this person is called in some hotels—not only heads the operation of the accounting department but is also closely affiliated with the operation and executive management. Many hotel owners and operators consider hotel accounting a highly specialized field and accordingly will entrust the affairs of their hotels only to accountants with considerable hotel experience. They feel that service operations and time-and-cost accounting are so different from other businesses that only a hotel accountant can successfully do the job.

If you plan a career in hotel accounting, specialize while at school in accounting before entering hotel work. Then, at least, you will have the requirements for a controllership. Working your way up is the next step.

Auditing

The auditor and the members of the auditing staff examine the accounts of the hotel and analyze them for the controller and manager. They check to make certain that all entries have been properly made; they look for errors. In general, they make a complete check of all monies of the hotel, all records, all accounts receivable and payable, and other matters of importance to the hotel's financial welfare.

Auditing positions require experience and education similar to that of the controller. As with the controller, promotion to auditor comes after many years of hotel experience. From auditor, the next step up is promotion to controller.

Computer Services

With the increasing use of computers in the industry, many accounting records are on data banks. Not all hotels are into this program; some small hotels choose not to computerize because the costs are not justified given the size of their operation.

Where computer usage is common, opportunities exist for openings as well as advanced positions in this field. While the science of computerization is a specialty, it is adapted for use in accounting according to the individual requirements of each accounting activity.

Larry Kant, accounting manager with the Management Group, a hotel management company in Chicago, feels computers are making an increasingly important contribution in the accounting department and other areas of hotel work:

> The advent of computers has eliminated the need for additional accounting staff that used to perform many of the routine bookkeeping chores. Balancing the ledgers—a job that used to take hours—can be completed in minutes with the right computer software. Extending inventories with a ten-key calculator and pen once took all day. This tedious chore is now over in less than an hour.

Accounts Payable

The accounts payable section is responsible for checking bills presented against shipping invoices and receipts. Upon finding that they check, this department then authorizes payment. In some hotels, accounts payable may draw checks and even have authority to sign and issue them.

A high school education is the minimum requirement for positions in accounts payable. Accounting, bookkeeping, and other business training are also required in many hotels. There are, however, openings here for beginners as clerks. These are filled mostly by young men and women who have completed high school.

Promotion from accounts payable to the auditing staff is possible, if you possess the proper education and business training. A college degree is highly desirable. Promotion is also possible to the credit, front office, and purchasing departments. From here too, you can go up the ladder to managership.

Accounts Receivable

The accounts receivable section lists payments and keeps records of all money received. All room, restaurant, and other charges are noted on records kept by this section. Payments made by guests against these charges are also recorded here.

The systems employed by accounts receivable departments differ from one hotel to the next. In some hotels, cashiers keep ledgers and do their own postings. In some hotels, all charge slips are forwarded to the accounts receivable department where entries are made. The system will vary, depending upon the size of the hotel and the system set up by the controller and auditor.

Opportunities for positions here correspond with those in accounts payable, and educational and training requirements are the same. Promotions possible in the accounts payable department are also possible in the accounts receivable department.

Payroll Department

Payment of all wages and salaries, maintenance of payroll records, the issuance of payroll checks, and the coordination of the payroll disbursements against wage and hour scales are the responsibility of the payroll department.

This department is headed by a paymaster, and promotion to this position is generally from accounting and credit departments, although some hotels may employ paymasters with experience in other hotels or businesses.

Assisting the paymaster are:

1. Payroll clerks, who make up and issue paychecks, compute wage and hour scales, keep records, check timekeeper reports, and perform other functions of the paymaster
2. Payroll cashiers, who issue checks to employees

3. Payroll conciliation clerks, who check the payroll bank account regularly to make certain that there is sufficient cash on deposit to meet all payroll accounts
4. Miscellaneous clerks

Educational and training requirements are general and somewhat similar to those of accounts payable and receivable. Studies in accounting, bookkeeping, statistics, hour and wage computation, personnel, and other affiliated courses are recommended.

Beginners are employed in this department since many functions are clerical. There are many opportunities for advancement, and considerable experience can be gained here.

Other Opportunities in Accounting

There are opportunities for employment in the accounting department other than those just described. Depending upon the size of the hotel, certain other duties and functions fall under the supervision of the controller or accounting department. The scope of the accounting department's job responsibilities varies from one hotel to the next.

Large hotels usually maintain statistical departments that correlate pertinent data helpful for the future operation of the hotel. The statistical department records such data as these:

1. Registrations—to ascertain where guests come from, the percentages of the geographical derivation of business, and other geographical data
2. Food and liquor purchases by brand, dish, age of customer, or other breakdowns deemed important by management
3. Age group and gender of guests
4. Returns from advertising and promotion campaigns
5. Time study and payroll information

This information is put to use by the purchasing department, the chef, the front office, advertising, sales and sales promotion, public relations, and other executive offices.

In many hotels, room and restaurant cashiers are supervised by the accounting department. Their duties are directed by the house treasurer or chief cashier. Cashiers perform all the duties their title implies. They receive payments, post charges, make change, and keep daily records.

Food control and purchasing control are, in some hotels, maintained by the accounting department; in other hotels they are part of the chef-steward's department or purchasing department. In large hotels, both of these departments, as well as the accounting department, may employ control and checking personnel who double-check and correlate each other's figures. Ultimately, all purchasing and chef-steward accounts are checked by the accounting department for errors and to make certain all records have been properly entered.

Hours of employment in the accounting department are generally eight hours a day, five or six days a week. Conditions of employment here are similar to regular office work.

Accounting work gives one an excellent background in hotel management and operation. The supervision of financial problems is an important function, and the controller of the hotel probably is more informed about its costs, problems, and other operations than any other person in the hotel. Every financial transaction passes before this person. Many a controller has become manager of a hotel. It is a logical promotion.

Educational requirements are high for employment here and particularly for advancement. A high school education, at least, is required for top positions. While there are openings as clerks available to beginners, most employees are required to have some form of bookkeeping or accounting background.

The average starting income in this department is about $6.25 to $10.00 per hour, depending on the job and its responsibilities. Jobs further up the ladder vary greatly in remuneration as there are many factors to be taken into consideration that are not common to all hotels. The rate of income will depend on the size of the department, hotel, and city; the responsibilities; the volume of business; and the kind of system set up by the controller. In general, earnings here are as good as, if not higher than, in most hotel departments. Sometimes meals are provided for managers or other members of the department.

The Credit Department

The credit department is responsible for authorizing charges made by guests, issuing credit cards, investigating the credit status of guests requesting credit cards, notifying guests of the acceptance or rejection of their credit applications, making adjustments on statements when incorrect charges have been posted, and keeping records and files of all credit transactions. Credit departments determine the credit limit of guests in most hotels, and indicate this on credit cards by key letters or numerals. Credit department people are required to follow up on delinquent accounts by writing letters or using other methods.

A credit manager heads this department, and the main functions are performed by assistants. Promotion to credit manager is usually from assistantships in the credit department. The position carries a great deal of responsibility with it, and candidates are very carefully considered. Seldom will an outsider without hotel experience be brought in. Hotels are frequently judged in hotel circles by the reliability of their credit methods and judgments. Certain hotels and chains have credit departments with such good reputations that credit cards they have issued are accepted by many other hotels as proof of good credit standing. Other credit cards are checked and approved for clearance.

In some of the larger hotels, the work of the assistant credit manager is aided by credit investigators. It is their duty to check accounts where credit has been overextended, investigate fraud, and, in general, oversee credit operations to prevent any criminal action against the hotel.

The credit department supervises operation of a ''guest history'' section for the sales department in some hotels. This section records the special requests and particular likes of guests. This information is cross-filed so that these special desires can be noted instantly against reservation or registration cards. A guest history might include such information as this: one guest asks for an extra-long bed; one guest likes four pillows; another guest always insists on southern exposure; one guest does not want to be higher than five floors above street level. The history also includes other pertinent information, such as the number of stays per year of each guest, the time of the year they check in, and so on. This information helps the sales department decide whom to favor in peak periods. To aid credit people in their work, the guest history section may also keep in this cross-file names of delinquent accounts and bad credit risks so that their arrivals or reservations are noted instantly by the credit department and preventive action can be taken.

Work hours in the credit department are based on the three-shift system. In large hotels, there is always someone on night duty. In the small hotels, the functions of this department are assumed by the general cashier or the owner-manager.

At least a high school education, and preferably college, is required in this department. Since there is much responsibility placed on the members of the credit office, most hotels are generally unwilling to employ people for credit work who have had no experience or education. When openings occur here, other employees of the hotel are considered, with education and background being important factors. From credit, promotions are made to front office or managerial staff.

Starting salaries in the credit department are about the same as in general accounting, depending on the size of the hotel and other factors. Regular increases augment the earnings here, and in some hotels, meals are provided managers and assistant managers of the credit departments.

THE PURCHASING DEPARTMENT

The purchasing department, while actually a back-of-the-house operation, is included in this front-of-the-house section since it performs a largely managerial function. The acumen of the purchasing manager and the efficiency of the purchasing department can make for profitable operation of a hotel.

Duties of the purchasing department include interviewing salespeople, placing orders for goods needed by all hotel departments, keeping records of all purchases and payments, drawing up and signing contracts and agreements for the purchase of all goods, comparing price and quality on all bids received, receiving and checking the quality and quantity of merchandise received on order, checking receipts and shipping invoices against accounts payable and forwarding such information to the accounting department, suggesting changes in the use of certain goods where costs can be saved or quality improved, and suggesting new products.

In some hotels, heads of both the housekeeping and the chef-steward department do their own interviewing of sales representatives, placing of orders, checking, and other functions of purchasing. This will depend largely upon the hotel and its size. The systems may vary accordingly.

The head of this department is the purchasing agent, or manager. This person supervises the functions and interviews, instructs, disciplines, and discharges employees in the purchasing department. Promotion to this position is usually from staff positions in the purchasing department. Occasionally, a hotel may employ as purchasing agent someone who has had considerable purchasing experience in other hotels or other businesses.

Experience in purchasing work, merchandising, and allied fields gives you an excellent background for purchasing work. Specialization in some particular phase of purchasing is sometimes also required as, for example, the purchasing of canned goods, office supplies, food, liquor, or linens. In some hotels, employees from other departments are considered for advancement to jobs in the purchasing department. Promotion will depend largely upon the individual, his or her education and experience, and the responsibility of the opening.

Purchasing checkers handle invoice control, examine incoming invoices to check errors, check invoices against purchasing department records and purchase orders, and verify the quality and quantity of all goods received. They notify the purchasing agent of vendors' compliance with all terms of purchasing orders and contracts.

A high school education is generally required for employment here, and college or hotel training courses are preferred. While there are such beginners' jobs as clerks, secretaries, and office helpers available in the purchasing department, opportunities for better jobs and promotions depend upon experience, ability, and training. Other fields in which experience can be gained are selling and estimating.

Most purchasing department employees work eight hours a day, five or six days a week. In general, office help will be on duty 9:00 A.M. to 5:00 P.M., while checkers or other helpers who are concerned with incoming shipments may work staggered or late hours in order to meet these shipments.

Since purchasing work is highly specialized, most hotels try to train staffs for this work and keep them as long as possible since they are not easily replaced. This is skilled work, and ability is generally equal to the length of time and experience spent in this work. While purchasing agents have been promoted to managerial positions, purchasing is generally considered to be a field in itself.

Starting salary can range from $5.00 to $10.00 an hour in the purchasing department. Much depends upon the job requirements or the individual's own experience and background. There are too many variable factors here to give a precise amount. The head of the department may not only receive a large salary but receive a bonus for savings made in purchasing or efficiency of operation. Sometimes meals may be furnished to certain members of this department.

CENTRAL FILES DEPARTMENT

A central files department is usually found only in large hotels, where files are most often kept on computer discs. In small hotels, its functions are absorbed by other departments, as the number of files doesn't warrant setting up a special section.

As large hotels have become older, their files have increased along with the years. And as these files have increased, it has been found impossible to have each department keep its own files. This would require more space for files than rooms.

As a result, this central file system was set up. It includes a central sorting and clearing center where all files are sorted, duplications weeded out, and central mailing lists set up. Today, much of this work is kept on computer. This might include general correspondence, bulletins, executive memorandums, contracts, guest files, and other information.

Added to these duties in some hotels have been those of general storekeeper and interoffice mailroom. The general storekeeper stocks and issues all office supplies. He or she keeps a constant check on inventory and sees to it that supplies are available for departmental use when needed.

The interoffice mailroom distributes all interoffice correspondence and handles the mailing of all hotel mail and packages.

This central mailing center not only saves the time of different departments but also helps keep one central control over all postal expenditures. In large hotels, postage is an expensive item.

The chief file clerk supervises the work of this department, and positions here include file clerks, mail personnel, storekeeper, and assistants. No special educational requirements are needed, although at least a high school education is preferred, and an understanding of computers is helpful. Many hotels give part-time employment here to students attending hotel training schools.

Promotion from the central files is possible to other departments of the hotel. The chief file clerk's position is a skilled position, since knowledge of filing systems and controls is necessary. The chief file clerk is appointed from other file clerks or other departments of the hotel. Occasionally, a hotel will employ as chief file clerk someone who has had filing experience in other hotels or businesses.

Remuneration is generally averaged at about $4.50 to $8.00 an hour, depending upon the hours, experience, and duties. But the central files department is the nerve center of the hotel and an excellent point from which to see the hotel in operation. Computerization is creating many new job opportunities here.

THE SECURITY DEPARTMENT

Hotel security departments range from a solitary employee in some small hotels to as many as twenty or more people in some of the larger hotels. Today's house officers and patrol personnel are primarily concerned with the protection of hotels guests and their property.

In most of the larger hotels, this department, under the supervision of an assistant manager in charge of protection (or chief house officer), operates quietly to safeguard hotel guests and property against theft or other crimes. Members of the department are stationed in public rooms, in the lobby, on banquet floors, and on the guest floors. When you realize that a large hotel is a city within a city with as many as three or four thousand guests in the house at one time and thousands of dollars worth of furnishings distributed throughout the building, you can see why potential criminals are likely to be attracted.

Technological advances have changed the way security systems operate. For example, in many hotels a room key is not used, and in its place a key card the size of a credit card is inserted in the room door to gain entrance. When a guest checks out of the hotel, the combination on the key card is reprogrammed.

In addition to their other duties, house officers are also trained to help distressed guests to their rooms, prevent disturbances in any part of the hotel, accompany cashiers, prevent annoyance of guests, and take charge in case of emergencies.

While educational requirements for this department are not specified, some house officers are college graduates, and many are former police officers. High school and college education, as well as hotel training courses, are helpful for promotion. Promotions are possible within the security department and also to other departments such as front office, credit, and management. There are several prominent hotel managers today who started as house officers and worked their way up. The ranks of house officers have often been tapped to fill openings for assistant managers on the floor.

In addition to house officers, the security department is composed of uniformed patrol personnel in large hotels. They regularly tour all guest floors, service floors, and public rooms of the hotels. In many hotels, they punch time clocks at stations along their tour. They inspect the premises continually to see that things are in order. Other patrol personnel are assigned to the receiving entrance to prevent loss of merchandise. They are also assigned to patrol work at conventions, banquets, and when large crowds are in any of the public rooms.

The security department may also house the lost and found section where all articles left in rooms by departing guests, or found elsewhere in the hotel, are kept for return to their proper owners. In some hotels, this function is handled by the housekeeping department. Here, too, reports are made of losses and are given to house officers.

Since security is a 24-hour operation, the three-shift system is generally employed by members of the department. Remuneration for employees of the security department averages about $250 to $450 weekly, depending on the size of the hotel, the hours, and the employee's experience. The executive positions in this department pay much more. Meals are sometimes furnished to the manager or other members of this department.

THE HUMAN RESOURCES DEPARTMENT

In large hotels, separate departments handle various aspects of personnel management duties. These duties include helping employees fill out application cards; keeping files on all employees; interviewing applicants for positions; advising applicants of their fitness for the various openings in the hotel; keeping time records;

investigating the references of all applicants; recording changes in employees' earnings, hours, jobs, education, training, home addresses, or telephone numbers; recording absences because of sickness or other reasons; noting merits, bonuses, disciplinary comments, recommendations from department heads, and causes for discharge. Personnel work also includes supplying references to other companies requesting information about previous employees, keeping lists of employees being considered for promotion, and supervising assignment and control of lockers. Personnel department members also analyze the various jobs in the hotel to determine special requirements or characteristics most needed to perform them well.

One key position in the department is that of timekeeper. The timekeeper is responsible to the paymaster. It is this person's duty to record the time of employees' arrivals and departures where there are no time clocks. He or she fills out time sheets, services the time clock, and performs other such duties. The timekeeper's reports are used by both personnel, for records, and by the paymaster, for computing the payroll.

Because a considerable number of hotel employees wear uniforms of one kind or another, most hotels provide lockers and dressing rooms where employees can change clothes. The personnel department locker crew regularly checks employee lockers, replacing locks when employees leave employment or are discharged and repairing lockers as needed. The crew also looks for articles of value left behind in the dressing room, or in unlocked lockers, and keeps track of them.

These personnel functions are supervised by the department head—the personnel director. While many personnel directors have worked their way up, today's hotels demand people with special education and training in personnel work. In most cases, a college education is required of applicants for this department. Many colleges today have special courses of study in personnel and human resources. In many instances, the personnel director participates in labor negotiations and may also supervise employee relations. In this case, specialized education and training is a must. Promotion from personnel director is straight up the ladder and may lead towards a managerial office.

Opportunities to enter personnel work are offered persons who have not had hotel experience, if they have completed educational and human resources work. In some instances, hotels will promote other workers to positions in the personnel department. Conversely, employees in the personnel department are eligible for promotion to other departments where openings exist. By the nature of their work in personnel analysis and job evaluation, personnel workers learn a great deal about hotel operation. The experience to be gained in this department is invaluable in starting a hotel career.

The average salary in this department is between $325 and $450 a week based on a regular eight-hour, five- or six-day week. Hours are generally regular, from 9:00 A.M. to 5:00 P.M.

THE BANQUET AND CATERING DEPARTMENT

Since the banquet department is primarily concerned with food and its service, one would think that this department naturally belongs in the back of the house. But since hotel banquet departments deal directly with guests or groups desiring space for conventions, meetings, luncheons, dinners, and other functions, the banquet department is listed here under front of the house.

In many hotels, banquet and catering functions make up a large portion of the profit from food operation. Group banquet business results in considerable revenue, not only from food and liquor sales, but from room rentals as well. A considerable amount of room business has been lost by hotels with inadequate banquet facilities.

Associations and business groups require a large amount of public space for exhibits and meetings at their annual conventions. They also need adequate ballroom and public space to accommodate the general luncheons, dinners, and meetings held for their membership. Accordingly, when officers of these groups plan their annual conventions, they are guided in their choice of a city and hotel by the size of the facilities available to accommodate their group. For this reason, cities that can provide adequate convention space—including large assembly halls for general meetings and enough small meeting rooms for divisional meetings—attract this large group business. Cities like New York, Atlantic City, Chicago, San Francisco, St. Louis, Miami, Miami Beach, Dallas, Fort Worth, Boston, Washington, and others even have special bureaus to solicit such group business. Because it results in extra billions of dollars in revenue for their hotels, restaurants, amusements, theaters, stores, and transportation facilities, cities vie for this convention business.

Such large group business results in concentrated service and good profits, and hotels seek to obtain as much of this business as possible. Banquet facilities and the operation of the banquet department figure largely in how much large group business an individual hotel receives.

Banquet Manager

The banquet manager, who supervises this department, is usually responsible to both the catering manager and the director of sales. This is so because the banquet manager fulfills two functions: one as a food manager and the other as a salesperson. In many hotels today, the trend is to make the banquet department, although it operates independently, part of the sales department.

All arrangements for banquets and other social functions are supervised by the banquet manager. This person directs the physical setups at all functions, draws up contracts and signs them, suggests or arranges for entertainment, and cooperates with all other departments involved in serving group business, such as the front office (rooms), sales department (who may have brought this business to the hotel), housekeeping, and others. The banquet manager is responsible for the efficient operation of all functions at the affair and must see to it that the hotel carries out its part of the bargain.

Banquet work is highly specialized and requires experience not only in planning menus and arranging meeting and convention setups but also in food costs and control. A banquet manager must know how to eliminate costly items from banquet menus without reducing the quality or appearance of the meal. He or she must know how to increase the sale of profit-bearing food and liquor and be able to sell his or her personality and ability to guests and committees.

The banquet manager can reach this position by starting at the bottom and learning each phase of food operation on the way up, or he or she can prepare to work in this department by taking special educational and culinary courses. In some hotels, promotion to banquet manager is made from the sales department, chef-steward's office, or managerial staff.

Banquet Staff

The size of the banquet staff depends upon the size of the hotel and its banquet operations. Their duties include selling space, scheduling events, keeping date books to avoid duplications of bookings, arranging for the listing of daily and weekly events on bulletin boards, suggesting and making up sample menus, arranging all functions for social events, setting up menus and programs, and servicing all functions.

Years ago, a young person interested in the banquet department apprenticed as a chef's helper, then became a kitchen assistant, next an under or assistant chef, and up the ladder to a *chef de potage,* or salad chef. The next step would be promotion to chef or banquet manager. Today's banquet people are trained personnel usually with training in food and hotel operation. Many excellent schools specialize in food and restaurant control and operation, so important to a banquet manager.

If you plan to enter the banquet department, you should therefore prepare by specializing at school in food and hotel courses. If this is not possible at the school or college you are attending, make arrangements to attend a school that has a course in hotel and restaurant operation.

After you have completed your education, start out in the chef-steward's department or as a beginner in the banquet department. The chef-steward's department is preferable because you will gain better groundwork in food preparation, cost, administration, and menu preparation here than in any other department. Many who have succeeded in the banquet field first started out as assistant waiters or waiters in hotel or outside restaurants. Food experience is important not only to success in the banquet department but to future success in any hotel career.

Banquet departments also include staffs of banquet waiters supervised by the banquet head waiter. These are waiters specially trained in banquet operations, which differ from regular table waiting. To aid the waiters, there are assistant waiters and housekeepers who set up tables, bars, and buffets before the waiters furnish them. The housekeepers clear out furnishings after each function.

Educational requirements for these positions are not rigid. If you cannot continue with advanced schooling, you can educate yourself in food operation by working in food departments. If you start as an assistant waiter or waiter, do not rely on this kind of on-the-job training exclusively. Once you have gotten your feet on the ground, make arrangements to augment your practical experience with courses in food costs, control, preparation, and operation. Also add courses in hotel management and operation. These will help round out your experience and facilitate your advancement. Many of the most successful hotel executives started out on the lower rungs of banquet operations. Opportunity is yours here, and your advancement will depend upon your own education, personality, ability, energy, ambition, and will to succeed.

Hours and Wages

The workweek in the banquet and catering departments has no set hours as in other departments; the schedule is staggered. Since functions take place in the evening, banquet employees are frequently asked to come in late and stay late. The hours will vary greatly even from day to day.

It is difficult to estimate the average earnings for waiters and other such personnel since their incomes depend upon the amount of business booked and the amount of tips they receive. Banquet work also can be very seasonal (weddings in the summer, for example), and earnings can fluctuate widely. Banquet employees average between $200 and $475 per week and more as they advance to assistant banquet managers. The top job here pays handsomely and is augmented by bonuses and percentages. Meals are sometimes provided for certain employees.

PUBLIC RELATIONS AND ADVERTISING

Public relations and advertising are specialties in their own rights. Most hotels fill openings in these departments with personnel who have had this kind of experience in other hotels or other fields. While advertising procedure and practices are fairly similar in hotels and in other businesses, public relations procedure for hotels is somewhat specialized.

The problems and tasks that confront persons in hotel public relations are as varied as the colors of the rainbow. On one day the public relations executive might be called upon to prepare a program for a technical education group; the next day he or she might be called upon to publicize a variety show in the hotel's main dining room. In the larger hotels, the director plans and supervises sales promotion activities in addition to supervising the public relations program.

Hotel public relations representatives are really executive assistants to top management. They are constantly called upon to represent and speak for the executive branch of the hotel. They must have a complete understanding of general hotel operation and policies. They must know what is required of each worker in the hotel and the hour and wage schedule of every department. Constantly called upon to supply facts and figures, public relations personnel must be equipped with essential knowledge of the hotel and its departments. A public relations executive—the title may be either manager or director—must have good judgment, experience, and at least a college education.

Public relations is an important profession in the hotel industry when you realize that a hotel sells primarily service, something so intangible that it must be measured solely in terms of public acceptance and recognition.

Advertising operations in a hotel are similar to those of any other business. The advertising manager or an outside agency prepares newspaper and magazine copy and suggests the appropriate media. The advertising manager in a hotel is also responsible for internal displays, such as those on elevators, in lobby easels, window displays, dining room table cards, and others. This person also supervises the printing of menus, programs for banquet functions, and all other printing, including stationery, business cards, billheads, ledger cards, and so forth.

Most advertising and public relations jobs go to persons with some experience in these fields. College education is usually required. However, there are opportunities to enter these professions as an apprentice and to educate yourself with on-the-job experience. This will depend purely upon the size of the hotel and the advertising and public relations staffs.

Remuneration varies with the hotel. Salaries can start at $275 weekly and range as high as $35,000–$55,000 yearly. Meals and, sometimes, lodging are provided. But there are many chances to go up the ladder here, which makes this an especially desirable department in which to gain hotel experience early in one's career.

THE SALES DEPARTMENT

Responsibility for selling space in public rooms and bringing large group business to the hotel belongs to the sales department. The volume of sales, and work, depends on the amount of space and other public facilities available.

Sales management has become an integral part of hotel operation and management. With sales in this industry now assuming a major position of importance, the sales director is regularly consulted regarding hotel policy and operation. In the larger chains, he or she is responsible only to the managing director or president and has complete authority and responsibility for front office, restaurant, banquet, and management policy.

This authority is easily understood when you stop to realize that sales is the department actively going after business. The sales department is in charge of all advertising and promotional expenditures; they determine which market should be exploited to realize the best results in room and food sales for the hotel. Therefore the sales department certainly should have control over the allotment of the merchandise it is selling.

Hotel sales personnel are essentially the same as sales forces in any business; their problems are the same. While there are no specified educational requirements, a high school education, at least, is preferred, and a college education is advantageous for future advancement. Completion of special courses in hotel management and operation will benefit persons interested in furthering their careers in sales work.

Sales departments differ with the various hotels. While one hotel may appoint one person as the director of sales or sales manager and label the rest of the staff as assistant sales managers, other hotels have given these assistant sales directors such titles as sales manager, convention manager, merchandise manager, and foreign sales manager. In hotels where the sales assistants have been given such titles, each salesperson specializes only in the type of business her or his title implies. While the convention manager goes after conventions, the merchandise manager goes after buyers and mercantile firms. However, the trend today in hotel sales work is away from this subdivision of departmental activities. Often certain business prospects lead to others, and it is not efficient to shift people according to the type of business.

Sales representatives, like credit people, are often given assistant manager titles since this aids them in their contacts. Many businesses have found it advantageous to appoint numerous vice-presidents in their sales departments for the same reason.

The sales director or manager has the duty of assigning leads or accounts to the various salespeople. Many salespeople receive percentage bonuses in addition to their salaries, and the sales manager must avoid favoritism in assigning accounts. The sales director must also work to maintain a good relationship with other departments in the hotel.

Close liaison between sales and banquet departments is required, for example. Since the major part of sales are those of public space, the work of both departments must be closely coordinated. To prevent duplicate bookings, one master entry book is usually kept. Once an event has been booked, the salesperson alone, or with the parties concerned present, arranges the setup, menu, and other details of the affair or function with the banquet manager.

The sales force also works very closely with the public relations department since sales promotion is one of this department's functions in many hotels. And even in those hotels where sales promotions (as in industry) are handled by a separate department, public relations work is generally called for with each group, including program advice and planning, press releases, speech writing, publicity, special events, photography arrangements, and other customary duties of public relations.

Beginners may enter the sales department directly from the outside, although front office clerks, credit people, accounting personnel, banquet representatives, bellhops, and others are often considered when openings occur in sales departments. Important characteristics sought by hotels in their salespeople are intelligence, good appearance, and the ability not only to sell but to get along with people.

Working hours are staggered. Many contacts are made at affairs or dinner parties, and sales reps frequently have to work evenings in order to develop business. Then, too, many prospects have their own business to occupy them in the daytime and are available for sales presentations only in the evenings. Along with irregular hours, hotel salespeople spend time on the road, contacting association officers and businesspeople at meetings and conventions in other cities.

One is usually promoted to managerial work from sales work. As the one who brings in the business, the salesperson has a following and therefore has a particular value, especially after many years of experience.

The average income received by salespeople is about $475 weekly and may run as high as $30,000 to $65,000 yearly in executive sales work. But it is difficult to draw an exact picture of earnings—you make your own way in sales work, and you also make your own salary. Many hotels give substantial bonuses for jobs well done.

In a carefully prepared and analytically thought-out message, Frank W. Berkman, former director of the Hotel Sales Management Association International, has put the picture into very clear focus. Mr. Berkman, who has been a well-known hotel executive most of his life and active in sales and hotel management for many years, writes:

> The hospitality industry today, more than ever before, offers unlimited opportunities for anyone seeking a challenging, stimulating, and rewarding career. In many countries throughout the world, as well as in numerous areas and provinces in the United States and Canada, the combined hospitality-tourism field is either the first or second largest industry in terms of business volume.
>
> Hotels, motels, and resorts have a dramatic impact on all sections of economic, social, and cultural life. The hospitality industry is totally "people-oriented," providing personal benefits to the countless millions of persons who use hotel/motel accommodations, facilities, and services around the globe.
>
> To encourage and further expand this use, whether it is business or pleasure-oriented, and to continually secure

profitable levels of room, food, and beverage sales—are all primary functions of sales and marketing.

There is a certain glamor and allure associated with hotel/motel sales promotion, advertising, publicity, and public relations. These include opportunities to meet famous and fascinating people from all walks of life, to travel, to entertain, as well as status and prestige, excellent industry advancement, and high salary potentials. Perhaps more significant are the unique opportunities hotel/motel sales and marketing can offer you in the fulfillment of your own very personal career wants and needs.

For example, ask yourself these questions: Would I enjoy the challenge of motivating people to purchase useful services or products, particularly by face-to-face selling? Would I particularly like the areas of business management and administration . . . of being in charge of an active, productive sales office? Do I especially seek out opportunities to use my creative abilities? Am I better suited to detail work—such as that involved with proper servicing after the sale is made?

If your answer to any, or a combination of, the above is "yes," then there certainly is a most profitable place for you in hotel/motel sales and marketing. What type of place? What specific job or position? The following offers a brief description of just some of the wide variety of challenging career positions in hotel sales and marketing.

Job Titles

Vice-President—Marketing: Establishes annual marketing program aimed at developing maximum business volume for rooms, food, beverages, and other sales; prepares sales goals and budgets; trains and develops sales personnel; and supervises and coordinates all related activities such as direct selling, advertising, publicity, and public relations.

Director of Sales: Administers, coordinates, and supervises sales department executives who are responsible for soliciting and servicing conventions, sales meetings, tours, and other groups requiring public space and room accommodations. Also creates and implements programs aimed at stimulating individual room, food, and beverage business.

Director of Advertising: Develops coordinated advertising campaigns and programs involving newspapers, magazines, radio and television, outdoor advertising, and direct mail. Works closely with advertising agencies in the creation and production of advertising and promotional literature.

Director of Public Relations: Responsible for developing positive programs directed at maximizing the hotel's image and its relations with the community, its employees, its guests, and the general public.

International Sales Manager: Coordinates activities specifically aimed at stimulating and developing both individual and group business from areas outside the country.

Tour and Agency Manager: Responsible for developing both group and individual business through personal contacts with travel agents, tour operators, transportation companies, and carrier representatives.

Convention Service Manager: Coordinates all hotel departments to assure maximum service to conventions and other groups once they are in the hotel, and is responsible for supervising all "in house" activities of the groups which involve hotel services.

Sales Representative: Directly contacts both repeat and new business prospects on a regularly established basis—through personal visits, telephone calls, and direct mail—for the specific purpose of booking a continuing flow of profitable business.

The need for qualified sales personnel is an ever-present one. New hotel/motel/resort building, the resulting increase in competition, and the ever-expanding market potentials—both domestic and international—all help make the experienced, professional sales and marketing executive one of the most sought-after employees in the hospitality industry. Yet the qualifications for careers such as those described in the preceding pages are not as technically demanding as you might think. The most important traits for a sales and marketing executive are positive personality, attitude, work habits, and relationships with others. Technical skills can be readily learned.

But because of the unique nature of the hospitality industry, there are a number of special qualities that are essential for those desiring to be successful in selling and servicing of its products.

First, empathy—the ability to put yourself in the other person's "shoes," such as when motivating a customer to buy by appealing to his or her specific needs and wants.

Second, initiative—the capacity of being a self-starter, to seek out and explore new business sources and potentials.

Third, creativity—the capacity to develop new, special, or unique marketing programs, attractions, or selling techniques, so that the benefits of your particular property stand out among all others.

While nothing can take the place of actual on-the-job experience, there are ample opportunities for those of you in high school and college, for example, to help build a proper foundation for your hotel career. Be sure to include in your curriculum courses in sales promotion, advertising, marketing, merchandising, tourism, motivation, communications, and public speaking. Actively participate in sales seminars and marketing workshops, such as those conducted by the Hotel Sales and Marketing Association

International (HSMAI) for both industry and colleges. And, HSMAI student membership is highly recommended as an extremely low-cost means of obtaining information on all facets of sales and marketing. HSMAI is located at 1300 L Street, N.W., Suite 800, Washington, D.C. 20005.

THE OPERATING MANAGEMENT

The resident manager is operating head of the hotel; this person supervises and directs all activities of the various departments. Generally, he or she is also on the executive board and is responsible only to the president or managing director of the hotel. The overall responsibility of the resident manager is to see to it that the guests are satisfied and that the hotel is operated as cleanly and as profitably as possible.

The resident manager has authority to make appointments or discharge any employee or department head for inefficiency, misconduct, or other reasons. He or she carries out the policies originated by the executive board or managing director and plans their execution by the various departments. At regularly held meetings of the department heads, the resident manager announces policy, schedules, and the execution of plans and discusses interdepartmental problems and conflicts. The resident manager also issues regular bulletins or notices to department heads and all other employees, notifying them of new policies, changes in operating schedules, hotel activities, and functions of interest.

As operating head of the hotel, the resident manager also participates in final negotiations with labor unions or employee groups after initial discussions and agreements have been prepared by the personnel department of the hotel.

The resident manager and the executive management must also be well experienced in the engineering problems of the hotel. Building and operating equipment problems, while under the direct surveillance of the chief engineer, are important management problems as well. Managers must be familiar with the most efficient types of machinery; they should understand furnaces, laundry machinery, kitchen machinery, and other equipment necessary to hotel operation. While they need not be expert on such matters, as engineering problems are handled by the engineering department, resident managers should nevertheless understand these problems well enough to prevent inefficiency in engineering practices. Many schools give special courses in hotel engineering to better acquaint management students with this important and expensive back-of-the-house problem.

Depending on the size of the hotel, there are any number of assistant and executive assistant managers to help with these duties. In general hotel practice, executive assistants are on duty each of the three shifts. Representing the manager, they are empowered to act officially in all situations coming to their attention.

In smaller hotels, the owner, who is usually the manager, may not only assume all the responsibilities of resident manager, but also those of the front office, credit, personnel, and other departments. Since the problems of small hotel operation are not considered the same as those of larger hotels, experience for resident manager openings in large hotels is best obtained in large hotel operations.

You generally attain the position of resident manager after many years of experience, and promotion to resident manager usually is made from the ranks of assistant managers, sales manager, credit director, controller, or other department heads. However, advancement to this top management position can be achieved by almost any employee in the hotel industry, including bellhops, accountants, chef-stewards, sales managers, and others, and one should constantly strive for self-improvement and progress. Either while employed in a hotel or before entering hotel work, one should complete courses in hotel management and operation given by many good schools and colleges. The position of resident manager is the aim of all interested in hotel careers and is available to all regardless of education and previous training. Remember, however, that the better your education and training, the better opportunity you will have to obtain your managership.

It is difficult to evaluate the earnings of hotel managers since they vary greatly, depending upon the size of the hotel, its operation, the policy of the executive branch, and other intangibles. According to a survey conducted for the American Hotel and Motel Association, earning of assistant and general hotel managers varied with the size of the hotel. Annual earnings for assistant managers in 1989 were $23,000–$40,000; general managers earned $37,000–$75,000. In certain cases earnings are augmented by bonuses and commissions paid on the basis of the business volume for certain periods. Apartments and meals are generally provided, also.

TOP MANAGEMENT

In the largest hotels, where operations rest in the hands of a resident manager and her or his staff, executive policy and control is in the hands of a higher executive branch. This may consist of a board of directors, an executive committee, or a managing director. This executive, or group of executives, represents ownership.

The executive, or group, formulates policy and supervises the actions of the resident manager. Although mostly concerned with financial matters and the accounting of profits or losses, the executive branch will take part in operational functions when called upon by the resident manager, or when conditions arise to make intervention necessary.

The executive management also arranges for financing when needed, decides on important changes in operations, approves investments for improvement or other reasons, hires top personnel, and, in general, supervises all top-level operations. Resident managers may be appointed to an executive post. Quite commonly, members of the executive board are chosen from banks, insurance companies, or other business groups that have financial or other interests in the hotel. There is no one plan of action that can take you to this level.

Remuneration varies. Some executives represent the owners or investors and receive their incomes outside of the hotel. Others, such as the managing director or president, receive their incomes directly from the hotel. Their earnings may consist of straight salaries augmented by bonuses for profitable operation or a percentage of hotel profits.

A famous hotel consultant advises newcomers as follows:

> In my experience in the hotel industry, I have become very sure that the industry offers a wide range of opportunities for an interesting and rewarding business career. To set aside, for the moment, the recognized fact that hotel employment offers the opportunity for a career generally without the boring aspects of many other careers, it also is true that numerous positions within the industry return to the individual monetary rewards ranging from fair to extremely good. Many years ago the rewards to be realized were limited but, particularly since the end of World War II, [management], department heads, and section leaders have enjoyed an upgrading of their wage scales to a point never anticipated prior to 1940.
>
> It is always a source of satisfaction to realize that so many of those in our industry who are in the better positions and

enjoying the most worthwhile fruits of their efforts are those who started in the hotels in a fairly menial capacity and through effort and attention have accomplished a steady climb to the pivotal positions which they now hold.

I would not want to underrate in any way the fine education which is being given to young men and women in the hotel schools of the many colleges and universities throughout the country. We see these graduates coming along side by side with the career worker to occupy the managerial posts.

With the reality of the new aspect of luxury motels throughout this country, and now being felt by the industry abroad, a new facet of innkeeping is opening up to those who are willing to embrace innkeeping as a life's career. All of us find interest and excitement in constantly being exposed to new people and new schedules, and this, being the daily fare in hotel operation, offsets the fact that we who have followed hoteldom as a career have, because of the very nature of the business, set aside the importance of a scheduled workday and a scheduled workweek.

For a young person with initiative, intelligence, and the desire to get ahead—and a basic liking for humanity—the hotel industry offers the golden opportunity for an interesting and successful career.

Indicative of the feeling, almost akin to that of show business, that reaches the inner part of people who have been engaged in hotel work is the statement once given us by Paul Grossinger, of Grossinger's, the famous New York state resort hotel.

The hotel industry today is certainly one to challenge the ability of any young person. Certainly no other business gives a person the opportunity of meeting so many various kinds of people, and no other business displays the human element as graphically.

A hotel is a world unto its own. We house and we feed people and also, in many instances, entertain them. We provide stopping areas, some as modest as a candy store, others as lavish as a series of shops operated by the best known names in the retail world. Certainly, an industry such as this is one to excite the imagination of the young. Most hotel people find that their business and social lives are greatly integrated. Most of us think that this is a benefit.

Financial gains in this industry are to the capable. Certainly, the basic concept of salary and wages in a hotel have gone up tremendously. Opportunity lies within the grasp of those who truly seek it.

Personally, I would not think of making my living in any other manner.

Milton Kutsher, of the famed Kutsher's Country Club in Monticello, New York, has this to say about a hotel career:

The modern resort hotel . . . exists only because it is today as necessary as home. A strong statement? Perhaps. But if one thinks about home as a place to relax and enjoy oneself, and then thinks about the modern resort hotel as a place to relax and enjoy oneself then we get a better idea of the place of the resort hotel in our mechanized culture. And from that, we get a better idea of the opportunities in today's modern resort industry.

The pace of the world of business and industry today makes vacations away from home a necessity. (Wasn't it a doctor who coined the phrase, "a change is as good as a vacation."?) Tensions can far more easily be released in an unfamiliar, sometimes exotic atmosphere than at home.

What's more, salaried people nowadays have both the income and the time to get away—unlike a few decades ago when resort hotels were neighborhoods for the employer and well-to-do management. Add to this the ever-increasing tendency of large companies to hold off-season conventions at resort hotels and one begins to see the rosy dawn of a new lease on life for the modern resort industry.

Of course, none of this would still amount to a hill of beans if it weren't all brought together with modern, high-speed transportation. What good the delights of the world if one has to spend all one's time getting there—not to mention getting back? So all of us in the modern resort industry owe a huge debt of gratitude to jets, super-trains, buses, super highways, and the people who are planning to speed up even further these super-speedy ways of getting anywhere and back.

And then, there's advertising. Which, for better or for worse, tells the world what we've got to offer. Together we have something for everyone. Something wonderful.

Something beautiful. Something exciting. Something relaxing. Together, the millions of dollars we in the resort industry spend on advertising have crystalized the vacationing habits of America so that today the resort world is healthy, flourishing, and still young.

Doesn't it follow, then, that opportunity for executive and management personnel has grown to an unprecedented degree? The person who wants—and trains himself or herself—to deal with the requirements of a demanding and sophisticated public will find that there is still gold in them thar hills, . . . for them who will dig for it.

CHAPTER 5

THE BACK OF THE HOUSE

The *back of the house* refers to those operations of the hotel that deal with housekeeping, food, and engineering, and which are seldom observed by guests. While restaurant operations involve direct contact with guests, they are so integral a part of food operations that they are described in this section. Although most top management and executive positions lie in the front of the house, the best sources for experience and the most opportunities for advancement lie in the back of the house.

Knowledge of food operation, control, and service is essential for profitable operation of a hotel. Because of increased costs of labor and materials, the minimum percentage of room occupancy at which hotels can be operated profitably has been rising in recent years. Successful food operations can be a major factor in profitable hotel operation.

Most knowledge about food is not obtained from books but only from actual experience and training. Complete your studies, prepare for hotel work in special hospitality schools, but also learn restaurant and food management and operation by actually working at it in the kitchen and the dining room. This experience will be a great advantage in furthering your hotel career.

One of the best ways to enter the hotel industry (and the restaurant field as well) is through the food and beverage areas, according to Brian Daly and Tony May, whose D-M Restaurant Corporation operates the internationally famous Rainbow Room and Rainbow Grill, both atop the sixty-fifth floor of Rockefeller Center, at 30 Rockefeller Plaza in New York City.

Writing for *Opportunities in the Hotel Industry,* Daly and May, whose company operates a full floor of private rooms that compete with New York's leading hotels for banquet business, said:

> We know of no other business that offers as many opportunities to neophytes just starting out on their careers than the food and beverage field. Opportunities, we mean, that are available to almost everyone who chooses this as a lifetime career, regardless of background, education, or environment.
>
> Many of today's top executives in food and beverage started at the very bottom of the ladder, some with more education and training than others, but all imbued with the same common element—a willingness to work hard and a desire to succeed.
>
> There are, in addition, more chances for beginners to get into the field, and perhaps these opportunities are greater than in many other fields because many of the starting positions are seemingly low—as assistant waiters, dishwashers, kitchen assistants, and similar "laboring" areas. But from these have come some of today's chefs, stewards, sales executives, and, yes, even managers of hotels and restaurants.
>
> This is not to belittle the importance of training and education. Not all chefs, stewards and other food and beverage executives started in lowly positions. European schools have long been turning out chefs of the highest order. To enter the business with a culinary degree or certificate from a European school, or hotel training course, is tantamount to entering the business world with an M.B.A. from Harvard. . . .

In addition, most of the schools and colleges in this country that turn out finished chefs and kitchen experts have the same aura of attainment as their European counterparts. In many instances, the "connections" made at these American schools of cuisine become important links later on in those hotels and restaurants where previous graduates have important spots.

We heartily recommend the food and beverage area of the industry, because it is first of all extremely challenging and interesting; second, because the preparation and serving of food is self-rewarding; third, because the field is an important one in the hotel and restaurant industry, if not in the entire economy; and last, because success in food and beverage is financially very rewarding at the top.

The food service industry is still the pioneer's frontier as a business venture and a profession. The last few decades have witnessed a constant expansion of food service in all of its segments—commercial restaurants, industrial and institutional food service, airline food service, and so on. Along with this constant expansion, new opportunities have been opened to thousands of men and women.

Looking into the future, it can be readily seen that the industry has not reached its limits. There are still many years ahead of us in physical and management development. The food service industry is still one of opportunity, perhaps more so than any other field. By the same token, the fact must be stressed that knowledge of the business details, of operations and management, is nowhere more required than in this industry.

In reality, a restaurant operator procures raw materials, manufactures the materials into a finished product, and finally places the product for sale on the market. Such a business process requires exacting knowledge because errors or ignorance can prove to be very costly. Perhaps that is why some of the most successful operators in the industry are those who have come up through the ranks. Regardless of formal education, knowledge and experience gained while working your way up through the ranks is extremely valuable and desirable.

However, promotion, even from the ranks, never is easy and simple. Competition for advanced positions is keen. As the industry matures, such competition will become even more pronounced. The axiom that advancement must be earned holds true in this industry as much as in any other.

THE FOOD AND LIQUOR DEPARTMENT

The activities of the food and liquor departments are generally supervised by one person. In smaller hotels, the manager or owner may personally supervise these operations. However, in larger institutions, the overseer of these departments is either an executive vice-president or a catering director. It is this executive's duty to supervise the food and liquor operations, to see that all foods purchased meet the requirements of the hotel, the menu, and the food cost policy. This person will also supervise the general service in these departments and dovetail operations with other departments where required. This executive must also keep close daily control over these operations so that at all times the operation and food costs are maintained at maximum operating efficiency and to the best advantage of the hotel.

One rises to this position only after many years of training in this field. A beginner cannot hope to aspire to this post except after many years of hard work and experience. As one of the top posts in a hotel organization, the pay here is quite high and often augmented by bonuses.

There are assistantships and office positions available in this department. The assistantships require almost as much experience as the top post, and appointments to the top post are often made from the ranks of assistants. Directly under the catering director are the chef-steward and wine steward, and, in some hotels, the banquet manager is partly responsible to the catering director for the food preparation and pricing of sales.

The Chef-Steward

The chef-steward is in charge of the preparation of all food sold in the dining rooms and through room service and banquets. He or she plans menus; purchases, prepares, and serves the food dishes; and supervises the various assistant chefs and other personnel in the department. The chef-steward usually is directly responsible to the catering manager. In some hotels, the chef is independent of the catering manager, and is sometimes assisted by a steward who makes purchases and supervises the noncooking or baking employees of the food department.

The purchase of food at hotels is usually a daily function. It would be impossible to store all the fresh vegetables, fruits, bakery products, meats, and fish it takes to provide the thousands of meals served daily by some of the large hotels.

As the catering manager and chef are both interested in food costs and control, menus are generally planned according to availability, season, and daily market quotations. In large operations, the saving of a fraction of a cent per dish can mean a good-sized profit. For this reason, the chef and catering manager try to base their menus on the best-priced seasonal items where they can make cuts and save, without impairing the quality of the food. Working as closely as they do, it is generally difficult to make up menus more than a day or two in advance.

The specific duties of the chef are discussed in detail in a later section of this chapter titled "The Kitchen."

Liquor and Food Controls

Since all three—the catering manager, the chef, and the wine steward—are responsible for operating their departments so that they show a profit, rigid food and liquor controls are observed.

The costs of preparing meals and drinks are figured down to the smallest fraction. When correlated with similar labor, overhead, and hidden costs analyses, profit or loss per portion can be shown.

Many restaurants and hotel food departments show a loss because of inefficient food controls. Hidden costs and wastes that do not show up on general cost figures can result in inefficient and unprofitable operation at the end of a fiscal period.

Some sort of checking system is employed in all hotel operations to control food and liquor orders. The systems vary with different hotels and types of employees. But some sort of control is necessary in order to prevent loss of revenue caused by inefficient billing or fraud. Checking is also important in compiling food statistics for use in analysis of food operations, costs, and profits. Checkers, responsible to the accounting department, perform this control function.

Room Service

Room service, which has been featured so often in Broadway and Hollywood comedies, must be carefully operated. Managed well, this can be a highly profitable food operation for a hotel, but it must be constantly promoted and conducted efficiently. In smaller hotels, room service is provided by bell persons or regular dining room attendants. In medium-sized and large hotels, room service is set up as a separate department.

There are many opportunities for employment here, including positions as waiters, assistant waiters, telephone order takers, assistant managers, and room service manager. The manager of this department is responsible for the efficient operation of the department and for the interviewing, disciplining, instructing, and discharging of the employees in room service.

In hotels where room service is provided as a separate service, the department is usually set up on a two-shift system, with the night shift eliminated. Occasionally, some hotels will stagger the day shifts so that service is provided until 1:00 or 2:00 A.M.

Room service positions can lead to positions of management in one of the dining rooms or the banquet department. Further steps up the ladder are to positions as banquet head-waiter, catering manager, and eventually managerial work.

There are opportunities to enter the room service department as an apprentice waiter. Most hotels insist on experience for waiting in room service, but some will employ people who have had no previous experience and train them. Positions in this department may lead to that of head of the room service department and so on up the ladder.

The Wine Steward

The sale of wines and beverages varies. In some areas, the sale of intoxicating beverages is forbidden. In others, local options are in existence whereby the laws may differ even from city to township. You will have to judge the setup according to local conditions.

In states where bottle clubs are common, hotel managements provide bar service, although the liquor is provided by the guests themselves.

In certain states, such as New York, hotel managements provide complete bar service, including service bars for use by waiters selling wines and beverages to the table trade.

Usually, the wine and beverage departments in larger hotels are supervised by the wine steward. An expert in the field, this person generally supervises the placing of orders, the storage, and the issuance of wines and liquors for use by guests. He or she is required to know good from bad vintage years, the proper care of wines and liquors, and the history of the profession and its products. The wine steward also supervises the work of the employees of the department, interviewing, instructing, disciplining, and discharging employees as required. He or she is responsible for seeing that wines and liquors are on hand in sufficient quantity and quality to meet all guest demands, that they are ordered according to demand, and that the department shows a profit from sales.

The position of wine steward is a highly honored one in a hotel and was originally handed down from generation to generation or given to one only after many long years of apprenticeship and experience as a wine steward's assistant. Today, promotion to this position is made from the ranks of assistant wine stewards or head bartenders. A great deal of specific experience is needed here, gained only from long, hard years of work and training in this department.

Of course, there are opportunities for beginners to enter the food and liquor departments. Openings exist for apprentices in the kitchen, for assistants and student waiters in the dining rooms, the banquet service, and room service departments, and for assistant bartenders and assistants to the wine steward in the liquor department.

For persons with experience, there are openings as bartenders in many hotels. Bartenders mix and serve alcoholic beverages and are required to know many, if not all, of the concoctions common to liquor service. Hotels with special house mixtures will train their bartenders in the mixing and serving of these drinks. Assisting the bartenders are the assistant bartenders, who chop ice, remove empty glasses or trash, bring in supplies, and set up ingredients for use by the bartender. From assistant bartender, the next promotion is to bartender or assistant to the wine steward.

From bartender, one usually advances to head bartender and then to wine steward.

Most bartenders are required to have previous experience, but assistant bartenders and cellarmen who have had no previous experience are given consideration. There are also numerous commercial bartenders' schools which offer courses of one to a few weeks at various tuition rates. If you consider a bartending school, it's wise to check its reputation with the institution where you want to work, and be sure the training you receive would be looked on favorably for employment. As with other hotel positions, a high school education at least is preferred, but beginners are trained in the duties and business of wines and liquors by the wine steward or an assistant in most large hotels. Smaller hotels generally have no openings for assistant bartenders, cellarmen, or wine stewards because of the small staff size.

The average bartender in the wine steward's department earns about $4.80 to $11.70 per hour, plus tips. These wages will depend largely on the size of the hotel and community and the responsibilities of the job, as well as on the amount of experience one has had in this field. Meals and uniforms are supplied by some hotels in addition to cash earnings.

The usual workweek is five or six days, eight hours a day.

The Kitchen

The preparation and serving of food has always been, and always will be, one of the most important and most skilled functions in any hotel, large or small. Since profit so frequently depends upon efficient and skillful operation of the food departments, the success or failure of a hotel depends in no small part upon the ability and experience of the chef-steward.

The best opportunities for entrance into the hotel field exist in the food department. There is actually a shortage of skilled trained executive chefs in this country. For every top hotel executive with experience and know-how of food operation, there are probably ten other hotel executives who lack such knowledge. Many hotel executives strongly recommend that the beginner consider this field before all others. In their opinion, knowledge of food is more important than almost anything else in the hotel business.

Large staffs of cooks who specialize in the preparation of different kinds of food are common in most large hotels. The head of the cooking staff is the chef-steward who plans the menus, orders the food, supervises the other cooks, institutes the style of cooking, and originates the recipes. This person is responsible for the ordering of sufficient food to meet all guest needs, proper preparation and serving of the food, and the operation of the department at a profit.

In some of the larger hotels, the chef-steward may be aided by a steward, who purchases the food and supervises the noncooking personnel. Other cooks in the kitchen may include a salad chef, cold meat chef, roast chef, sauce chef, dessert chef, and so on. There may also be butchers, bakers, and pastry chefs. The specialization will depend upon the size of the staff. In addition, there are helpers, assistants, and apprentices.

One of the most important jobs in the hotel field, the chef position is sought by most who enter the cooking field. It is reached generally only after many years of experience. While most chefs or underchefs are employed only upon the basis of their previous experience, one can enter this field with little experience and gain on-the-job training.

At least two or three years of apprenticeship in a large-staffed hotel kitchen is necessary in order to become a cook. Many hotels require additional years of training and experience as assistant chef in order to be considered for the job of chef. The years of training will vary somewhat, depending upon the size of the hotel and your own ability and talents for this profession. To help you in this career, numerous schools have instituted courses in this work, and for your convenience these classes are often given day and night. See appendix B for a list of schools and colleges providing such training.

Many hotels today are increasing their apprentice-chef training programs, and many more opportunities are being made available for persons interested in this work as a career. Not only do you learn a trade here, but also you get a solid background in one of the most important subjects needed for top executive management.

While most cooks are on a 40- to 48-hour week, these hours vary, and some cooks will even work as many as 70 hours a week. The hours, schedules, and times will vary with the hotels and communities.

The average weekly wage for most cooks is between $240–$400 a week for staff cooks, while executive chefs earn about $30,000–$75,000 per year, and more, in average hotels. In the large hotels, executive chefs make a great deal more, and it is difficult to estimate their earnings. They also receive bonuses for instituting savings on food costs. In addition to cash earnings, cooks, chefs, and other department members usually receive one or more meals daily.

Restaurant Operation

While restaurant operations in large hotels are *supervised* by the catering manger, they are *carried out* by their managers. In charge of each restaurant in the hotel is the restaurant manager, whose duties include the interviewing, instructing, disciplining, and discharging of employees; keeping records; handling customers' complaints; sometimes preparing menus or making suggestions for menu items; and supervising all the various activities that are required to make the restaurant efficient and attractive.

In some smaller hotels, a restaurant manager may work very closely with the chef in preparing the menus and purchasing the food. A thorough knowledge of preparing, storing, and purchasing food, as well as food cost accounting, menu preparation, and checking, is helpful in this work. In addition, the manager must be familiar with sanitary practices and local regulations. He or she also supervises and assigns duties to employees, seeing that no favoritism is shown any particular member of the department.

Minimum experience required for restaurant management ranges from one to five years, depending on the size of the hotel. The larger hotels may not only require the longer experience, but also may assign prospective restaurant managers to other duties in the food and restaurant department in order to familiarize them with hotel operations before appointing them to the managership.

The minimum education requirement is high school, and your career will be furthered if you have college and/or food and restaurant-management training at an accredited school. While most hotels and large restaurant chains start college-trained persons as assistant managers today, giving them courses of instruction before they promote them to managerships, there are still opportunities for waiters, cooks, assistant waiters, and others to work their way up to manager of a hotel restaurant.

The captain is an assistant to the manager and not only assists in managerial duties, but also conducts people to tables, assigns waiters to stations, and sees that the guests are seated at waiting stations in rotation so that not all are seated in any one station to the disadvantage of other waiters.

In addition to the restaurant manager and captain, hotel restaurants have staffs consisting of waiters, assistant waiters, cashiers, and assistants. Besides taking guest orders and serving food and liquor to tables, waiters also set tables, sometimes collect payment, make out checks, arrange setups, help bus when busy, and perform other chores.

While most hotels employ as waiters only men and women who have had experience in other hotels or restaurants or who have had experience waiting on tables, many hotels today are training their assistant waiters for promotion to jobs as waiters as openings occur. Some hotels are even engaging persons with no previous waiting experience in order to train them in the hotel's own system.

In this field of hotel operation hours of work and earnings vary greatly. Hours depend upon local hours of service, working conditions, and many other factors. Earnings are based on tips as well as salaries, and meals are usually provided also.

THE HOUSEKEEPING DEPARTMENT

Much of the reputation of a hotel lies in the hands of its housekeeper and the housekeeping staff. The most important thing sought by the average hotel guest is a clean, neat, attractive, cheerful, comfortable room. He or she also wants to see clean, neat halls and public rooms. An inefficient housekeeper can ruin a hotel's reputation almost overnight. If a hotel is to succeed, its standards must be kept high.

Heading the housekeeping department is the executive or head housekeeper. It is this person's responsibility to see that halls, rooms, and furnishings are clean and attractive. In larger hotels where housekeeping staffs are also large, the executive housekeeper also has these duties: assisting in or making purchases of supplies for the department; interviewing, disciplining, instructing, and discharging employees in the department; keeping employee and housekeeping records; making regular reports to the manager of conditions, repairs, improvements, employee problems, expenditures, and suggestions; keeping inventories; and making out the department payroll. Frequently a housekeeper, if skilled, will create, or supervise the creation of, new schemes of interior decoration.

In large hotels, the housekeeping staff may include linen room attendants, assistant housekeepers, floor supervisors, housekeepers, furniture polishers, wall and window washers, seamsters and seamstresses, upholsterers, painters, cabinetmakers, and others skilled in housekeeping repair and maintenance.

Promotion to executive housekeeper is usually made from the housekeeping staff or by employing persons with experience in other hotels. Frequently, inexperienced persons are employed as assistants to floor supervisors and given training in their work. While previous training and experience are usually preferred for executive work in the housekeeping department, many have risen to top positions here from lesser jobs. Excellent training courses for housekeeping jobs are given by many high schools and vocational training programs throughout the country. These courses should be of great help in entering this field.

Openings as maid, housekeeper, supervisor, and other jobs in the housekeeping department are available to persons with little or no experience, and application should be made to the executive housekeeper.

While lesser jobs in the housekeeping field often do not pay well, they are advantageous in that they are available to persons with little or no previous experience. To persons with the ambition and ability to succeed, these jobs offer opportunity to advance, since the rate of turnover in the housekeeping field is rather high.

Earnings of executive housekeepers average $350–$450 weekly in small hotels to $450–$900 weekly in large hotels. In some hotels, executive housekeepers earn much more. Meals and lodging are quite often given in addition to cash earnings.

CHAPTER 6

OTHER DEPARTMENTS

There are many other hotel jobs in addition to those specialized trades we have previously described. These other jobs, while important to the hotel's operation, are not hotel trades as such and do not require specific hotel experience. Among these other departments are engineering, telephone, laundry, valet, medical, and dental.

The hotel's water, heat, and other physical facilities are operated by the engineering department. Its size will correspond with the size of the hotel. In a large hotel, the engineering department will include boiler-room attendants, carpenters, electricians, engine-room attendants, maintenance engineers, plumbers, painters, compression workers, and others. Required experience for these positions depends upon the job to be filled. Hotel experience is not a prime factor in employment here. A background in a trade is more important than previous hotel work.

If you have a skill or trade that equips you for work in the engineering department of a hotel, talk to the chief engineer or to the personnel director for information about openings in your classification. Hours and remuneration will vary with the nature of your work and the size of the hotel organization.

There are also openings for telephone operators, laundry help, and valets (pressing and tailoring) in those hotels where these services are provided. Inquiry concerning openings, hours, and remuneration should be made to the personnel director.

In some large hotels, medical and dental services are available on the premises for the convenience of guests and for emergencies among the employees. These openings are filled from regular medical and dental channels. Medical or dental clinics can use nurses, receptionists, and secretaries. Make inquiry directly at the clinic or office.

In addition, there are numerous secretarial, typing, computer operating, reception and other jobs in hotels. Inquiries for these should be made at the personnel office.

THE LODGING INDUSTRY—ROOM TO GROW IN

Chairman and chief executive officer of Westin Hotels, Harry Mullikin, said that choosing a career is

> . . . at one and the same time, a thrilling, yet often frightening, decision for today's young men and women.
>
> In an age so dominated by technology, many young people find it increasingly difficult to identify career opportunities that meet their desires to become truly involved and contribute to our society.
>
> There is an industry that offers a wider spectrum of career opportunities—an industry that demands of its career professionals a commitment to involvement and rewards that commitment with the satisfaction of seeing things happen.
>
> That industry is the lodging industry—a complex, dynamic industry facing growth projections unequalled in today's economy. That growth factor means, of course, more and more career opportunities that are made even more attractive by the very variety of professional openings in

such fields as marketing, food and beverage, fiscal management, advertising and public relations, property management, and market research, to mention but a few.

At this writing, the hospitality industry projects its needs for management personnel in the scale of tens of thousands reflecting very real and challenging opportunities for men and women who seek the chance to innovate, to "imagineer," to bring to their working assignments the desire to participate and to contribute.

Perhaps of even more interest to the man and woman seeking a rewarding career is the basic fact of our industry; it is a people-oriented industry affording its employees the day-to-day opportunity to interact with the traveling public. There are few industries where creativity, innovation, and imagination are more welcome! There are few industries that can offer, as ours, the immediate response of the marketplace to new ideas.

Like any field of endeavor, the lodging industry is a demanding one requiring acceptance of long, and sometimes unusual, hours. It asks of management people a sense of responsibility and obligation to its guests that is uniquely different from other industries. And it is a changing industry requiring continuing study if it is to properly fulfill demands of the marketplace today and tomorrow.

Today, so many of our young people are seeking careers that provide the opportunity for self-expression, the challenge of creativity and innovation, the reward that comes from seeing one's ideas win acceptance and approval. The lodging industry offers that opportunity, that challenge, that reward.

SOURCES OF ADDITIONAL INFORMATION

There are hundreds of books and other publications that provide valuable reading in the hospitality field. We have listed several publications that provide general career and educational information. Your public library will have other publications covering current trends in the hotel and motel industry. Other good sources of information are the professional organizations serving this field. We have listed several of the most important of these organizations.

Publications

The Cornell Hotel and Restaurant Administration Quarterly. The School of Hotel Administration, Cornell University, Ithaca, New York 14853.

The Guide to Hospitality and Tourism Education. The Council on Hotel, Restaurant, and Institutional Education (CHRIE), 1200 17th Street, N.W., Washington, D.C. 20036.

Outlook 2000. Bulletin #2352, Bureau of Labor Statistics, Publication Sales Center, P.O. Box 2145, Chicago, IL 60690.

Organizations

American Hotel and Motel Association, 1201 New York Avenue, N.W., Suite 600, Washington, D.C. 20005.

Council on Hotel, Restaurant, and Institutional Education (CHRIE), 1200 17th Street, N.W., Washington, D.C. 20036.

Hotel Sales and Marketing Association International, 1300 L Street, N.W., Suite 800, Washington, D.C. 20005.

National Association of Black Hospitality Professionals, P.O. Box 5443, Plainfield, New Jersey 07060.

The Educational Foundation of the National Restaurant Association, 250 S. Wacker Drive, Suite 1400, Chicago, Illinois 60606.

SCHOOLS OFFERING HOTEL AND OTHER HOSPITALITY TRAINING

The following is a list of colleges and universities that offer an associate's or bachelor's degree in hotel, restaurant, and institutional management or food service administration. A number of schools offer either a master's in business administration or a master's degree in hotel, restaurant, and institutional management; these programs are marked with the letter *M* to the left of their listing. The list was prepared by the Educational Foundation of the National Restaurant Association, 250 S. Wacker Drive, Suite 1400, Chicago, Illinois 60606.

Although this listing is one of the most comprehensive available, undoubtedly schools were missed. You should consult with your local libraries and guidance counselors when you are ready to make a choice to be sure you have considered all the schools available at the time.

Four-Year Colleges and Universities

Alabama

Auburn University
  Hotel, Restaurant Management
    Program
  School of Human Sciences
  328 Spidel Hall
  Auburn, Alabama 36849-5605

Tuskegee University
  Hospitality Management
    Program
  Department of Home
    Economics
  Tuskegee, Alabama 36088

University of Alabama
  Restaurant and Hospitality
    Management Program
  College of Human
    Environmental Sciences
  P.O. Box 870158
  Tuscaloosa, Alabama 35487

Alaska

Alaska Pacific University
  Travel and Hospitality
    Management
  School of Management
  4101 University Drive
  Anchorage, Alaska 99508

University of Alaska
  Travel Industry Management
    Program
  School of Management
  Fairbanks, Alaska 99709

Arizona

Northern Arizona University
  School of Hotel and Restaurant
    Management
  NAU Box 5638
  Flagstaff, Arizona 86011

Arkansas

Arkansas Tech University
  Hotel and Restaurant
    Management
  School of Business
  Corley Building
  Russellville, Arkansas 72801

University of Arkansas at Pine
    Bluff
  Foodservice, Restaurant
    Management
  Home Economics Department
  P.O. Box 4128
  Pine Bluff, Arkansas 71601

California

California State Polytechnic
    University
  Center for Hospitality
    Management
  3801 West Temple Avenue
  Pomona, California 91768

California State University, Chico
  Foodservice Administration
  School of Home Economics
  117 Glenn Hall
  Chico, California 95929-0002

California State University, Long
  Beach
  Food Administration
  Department of Home
    Economics
  1250 Bellflower Boulevard
  Long Beach, California 90840

Chapman College
  Hotel, Restaurant and Tourism
    Management
  Orange, California 92666

Golden Gate University
  M Hotel, Restaurant and
    Tourism Management
  College of Business
    Administration
  536 Mission Street
  San Francisco, California
    94105

Loma Linda University
  Food Systems Management
  Department of Nutrition and
    Dietetics
  Loma Linda, California 92354

San Jose State University
  Foodservice Management
  Department of Nutrition and
    Food Science
  One Washington Square
  San Jose, California 95192

United States International
    University
  School of Hospitality
    Management
  10455 Pomerado Road
  San Diego, California 92131

University of San Francisco
  Hospitality Management
  School of Business
  Ignatian Heights
  San Francisco, California
    94117

Colorado

Colorado State University
  Restaurant Management
  Department of Food Science
    and Human Nutrition
  Fort Collins, Colorado 80523

Metropolitan State College
  Hospitality, Meeting, Travel
    Administration
  1006 11th Street, Box 60
  Denver, Colorado 80204

University of Denver
  School of Hotel and Restaurant
    Management
  2030 East Evans Avenue
  Denver, Colorado 80208

Connecticut

University of New Haven
  M School of Hotel, Restaurant
    and Tourism Administration
  300 Orange Avenue
  West Haven, Connecticut
    06516

Delaware

Delaware State College
  Hotel and Restaurant
    Management
  Department of Home
    Economics
  1200 DuPont Highway
  Dover, Delaware 19901

University of Delaware
  Hotel, Restaurant and
    Institutional Management
  College of Human Resources
  Alison Hall
  Newark, Delaware 19716

District of Columbia

Howard University
  Hotel, Motel Management
  School of Business
  2600 6th Street, N.W.
  Washington, D.C. 20059

Florida

Bethune-Cookman College
  Hospitality Management
    Program
  Division of Business
  640 Second Avenue
  Daytona Beach, Florida 32115

College of Boca Raton
  Hotel, Restaurant and Tourism
    Management
  3601 North Military Trail
  Boca Raton, Florida 33431

Florida International University
  M School of Hospitality
    Management
  15101 Biscayne Boulevard
  North Miami, Florida 33181

Florida State University
  Hospitality Administration
  School of Business
  225 William Johnston Building
  Tallahassee, Florida 32306

Saint Leo College
  Restaurant and Hotel
    Management
  State Road 52, P.O. Box 2067
  Saint Leo, Florida 33574

St. Thomas University
  Tourism and Hospitality
    Management
  School of Business,
    Economics, Sports and
    Tourism
  16400 NW 32nd Avenue
  Miami, Florida 33138

University of Central Florida
  Hospitality Management
    Department
  College of Business
  CEBA II #409
  Orlando, Florida 32816

Webber College
  Hotel and Restaurant
    Management
  1201 Alternate Highway 27
    South
  Babson Park, Florida 33827

Georgia

Georgia Southern University
Restaurant, Hotel, and
Institutional Administration
Division of Home Economics
Landrum Box 8034
Statesboro, Georgia 30458

Georgia State University
Hotel, Restaurant and Travel
Administration
College of Public and Urban
Affairs
University Plaza
Atlanta, Georgia 30303

Morris Brown College
Hospitality Administration
643 Martin Luther King Drive
Atlanta, Georgia 30314

University of Georgia
Hotel and Restaurant
Administration
College of Home Economics
Dawson Hall
Athens, Georgia 30602

Hawaii

Brigham Young
University—Hawaii
Hospitality Management
Business Division
55-220 Kulanui
Laie, Hawaii 96762

Hawaii Pacific College
Travel Industry Management
1188 Fort Street
Honolulu, Hawaii 96813

University of Hawaii—Manoa
School of Travel Industry
Management
2560 Campus Road, George
Hall
Honolulu, Hawaii 96822

Illinois

Chicago State University
Hotel and Restaurant
Management
College of Business and
Administration
95th Street at King Drive
Chicago, Illinois 60628

Eastern Illinois University
Hospitality Services Program
School of Home Economics
109 Klehm Hall, S.H.E.
Charleston, Illinois 61920

Kendall College
Hospitality Management
2408 Orrington Avenue
Evanston, Illinois 60201

Northern Illinois University
Food Science
Department of Human and
Family Resources
DeKalb, Illinois 60115

Roosevelt University
Hospitality Management
430 South Michigan Avenue
Chicago, Illinois 60605

Rosary College
  Foodservice Management
  7900 West Division Street
  River Forest, Illinois 60305

Southern Illinois University
  Food and Lodging Systems
    Management
  College of Agriculture
  Quigley Hall, Room 209
  Carbondale, Illinois 62901

University of Illinois
  M Hospitality Management
  School of Human Resources
    and Family Studies
  363 Bevier Hall, 905 South
    Goodwin Avenue
  Urbana, Illinois 61801

Western Illinois University
  Foodservice and Lodging
    Management Program
  Department of Home
    Economics
  204 Knoblauch Hall
  Macomb, Illinois 61455

Indiana

Ball State University
  Foodservice Management
  Home Economics Department
  Practical Arts Building
  Muncie, Indiana 47306

Purdue University
  Department of Restaurant,
    Hotel and Institutional
    Management
  M School of Consumer and
    Family Sciences
  106 Stone Hall
  West Lafayette, Indiana 47907

Iowa

Iowa State University
  M Hotel, Restaurant, and
    Institution Management
  College of Family and
    Consumer Sciences
  11 MacKay Hall
  Ames, Iowa 50011

Kansas

Kansas State University
  Hotel and Restaurant
    Management
  College of Human Ecology
  Justin Hall
  Manhattan, Kansas 66506

Kentucky

Morehead State University
  Hotel, Restaurant and
    Institutional Management
  Department of Home
    Economics
  UPO Box 889
  Morehead, Kentucky 40351

Transylvania University
  Hotel, Restaurant, and Tourism
    Administration
  Division of Business
    Administration
  300 North Broadway
  Lexington, Kentucky 40508

University of Kentucky
  Restaurant Management
  College of Home Economics
  210 Erikson Hall
  Lexington, Kentucky 40506

Western Kentucky University
  Hotel, Motel, Restaurant
    Management
  Department of Home
    Economics and Family
    Living
  Academic Complex
  Bowling Green, Kentucky
    42101

Louisiana

Grambling State University
  Hotel and Restaurant
    Management
  P.O. Box 882
  Grambling, Louisiana 71245

University of New Orleans
  School of Hotel, Restaurant
    and Tourism Administration
  College of Business
  New Orleans, Louisiana 70148

University of Southwestern
    Louisiana
  Restaurant Administration
  School of Home Economics
  USL P.O. Box 40399, 200
    East University
  Lafayette, Louisiana 70504

Maryland

University of Maryland
  Food Systems
  College of Human Ecology
  College Park, Maryland 20742

University of Maryland—Eastern
    Shore
  Hotel and Restaurant
    Management
  School of Professional Studies
  Somerset Hall, Room 409
  Princess Anne, Maryland
    21853

Massachusetts

Boston University
  Hotel and Food Administration
  Metropolitan College
  808 Commonwealth Avenue
  Boston, Massachusetts 02215

Framingham State College
  Foodservice Systems
    Management
  Department of Home
    Economics
  State Street, Hemenway Hall
  Framingham, Massachusetts
    01701

University of Massachussetts
   M Department of Hotel,
   Restaurant and Travel
   Administration
   College of Food and Natural
   Resources
   Flint Laboratory, Room 107
   Amherst, Massachusetts 01003

Michigan

Central Michigan University
   Marketing and Hospitality
   Services Administration
   College of Business
   100 Smith Hall
   Mt. Pleasant, Michigan 48859

Davenport College of Business
   Restaurant and Lodging
   Management Program
   415 East Fulton
   Grand Rapids, Michigan 49507

Eastern Michigan University
   Hospitality Management
   Human Environmental and
   Consumer Resources
   202 Roosevelt Hall
   Ypsilanti, Michigan 48197

Ferris State University
   Hospitality Management
   School of Business
   901 South State Street
   Big Rapids, Michigan 49307

Grand Valley State University
   Hospitality and Tourism
   Management
   Science and Mathematics
   Division
   AuSable 111
   Allendale, Michigan 49401

Michigan State University
   M School of Hotel, Restaurant
   and Institutional
   Management
   425 Eppley Center
   East Lansing, Michigan
   48824-1121

Northern Michigan University
   Restaurant and Foodservice
   Management
   School of Technology and
   Applied Science
   Jacobetti Center, Route 550
   Marquette, Michigan 49855

Siena Heights College
   Hotel, Restaurant and
   Institutional Management
   1247 East Siena Heights Drive
   Adrian, Michigan 49221

Minnesota

Mankato State University
   Food and Nutrition
   Home Economics Department
   P.O. Box 8400, MSU Box 44
   Mankato, Minnesota
   56002-8400

Moorhead State University
Hotel, Motel, Restaurant
Management
Department of Business
Administration
Moorhead, Minnesota 56560

Southwest State University
Hotel and Restaurant
Administration Program
Department of Hospitality,
Marketing, and Agribusiness
Lecture Center 101
Marshall, Minnesota 56258

Mississippi

University of Southern
Mississippi
Hotel, Restaurant and Tourism
Management
College of Health and Human
Sciences
S.S. Box 5035
Hattiesburg, Mississippi 39406

Missouri

Central Missouri State University
Hotel and Restaurant
Administration
Home Economics Department
Grinstead 250
Warrensburg, Missouri 64093

Southwest Missouri State
University
Hospitality and Restaurant
Administration
College of Health and Applied
Science
901 South National
Springfield, Missouri 65804

University of Missouri
Hotel and Restaurant
Management Program
Department of Food Science
and Nutrition
122 Eckles Hall
Columbia, Missouri 65211

Nebraska

University of Nebraska
Foodservice Management
Department of Home
Economics
316 Ruth Leverton
Lincoln, Nebraska 68583

University of Nebraska
Restaurant Management
School of Human Nutrition
60th and Dodge
Omaha, Nebraska 68182-0214

Nevada

Sierra Nevada College—Lake
Tahoe
Hotel, Restaurant and Resort
Management
P.O. Box 4269, 800 College
Drive
Incline Village, Nevada 89450

University of Nevada—Las Vegas
  M College of Hotel
    Administration
  4505 Maryland Parkway
  Las Vegas, Nevada 89154-6013

New Hampshire

New Hampshire College
  Hotel, Restaurant Management
    and Culinary Arts
  2500 North River Road
  Manchester, New Hampshire
    03104

University of New Hampshire
  Hotel Administration
  School of Business and
    Economics
  McConnell Hall
  Durham, New Hampshire
    03824

New Jersey

Fairleigh Dickinson University
  M School of Hotel,
    Restaurant, and Tourism
    Management
  Hesslein Building
  Rutherford, New Jersey 07070

Montclair State College
  Foodservice Management
  Home Economics Department
  Upper Montclair, New Jersey
    07043

New Mexico

New Mexico State University
  Hospitality and Tourism
    Services Program
  College of Agriculture and
    Home Economics
  P.O. Box 30003, Department
    3HTS
  Las Cruces, New Mexico
    88003

New York

Buffalo State College
  Food Systems Management
  Nutrition and Food Science
    Department
  1300 Elmwood Avenue,
    Caudell 106
  Buffalo, New York 14222

City University of New York
  Hotel and Restaurant
    Management Department
  New York City Technical
    College
  300 Jay Street
  Brooklyn, New York 11201

Cornell University
  M School of Hotel
    Administration
  Statler Hall
  Ithaca, New York 14853

Marymount College
  Foods for Business and
    Industry
  Department of Human Ecology
  Box 1375
  Tarrytown, New York 10591

New York Institute of Technology
School of Hotel, Restaurant
Administrration
Northern Boulevard
Old Westbury, New York
11568-9998

New York University
M Center for Food and Hotel
Management
East Building, Room 537
239 Greene Street
New York, New York 10003

Niagara University
Institute of Travel, Hotel and
Restaurant Administration
Niagara University, New York
14109

Rochester Institute of Technology
M School of Food, Hotel,
Tourism Management
1 Lomb Memorial Drive, P.O.
Box 9887
Rochester, New York 14623

State University of New York at
Oneonta
Food and Business
Department of Home
Economics
Oneonta, New York 13820-4015

State University of New York at
Plattsburgh
Hotel and Restaurant
Management
Center for Human Resources
Draper Avenue, Ward Hall
Plattsburgh, New York 12901

Syracuse University
Food Systems Management
College for Human
Development
112 Slocum Hall
Syracuse, New York 13244

North Carolina

Appalachian State University
Hospitality Management
Program
College of Business
Boone, North Carolina 28608

Barber-Scotia College
Hotel, Restaurant Management
145 Cabarrus Avenue West
Concord, North Carolina 28025

East Carolina University
Hospitality Management
School of Home Economics
Greenville, North Carolina
27858

North Carolina Central University
Institutional Management
Department of Home
Economics
P.O. Box 19615
Durham, North Carolina 27707

North Carolina Wesleyan College
Foodservice and Hotel
Management
Wesleyan Station
Rocky Mount, North Carolina
27803

North Dakota

North Dakota State University
Hotel, Motel and Restaurant
Management
College of Home Economics
Fargo, North Dakota 58105

Ohio

Ashland University
Hotel, Restaurant Program
School of Business
401 College Avenue
Ashland, Ohio 44805

Bowling Green State University
Restaurant and Institutional
Foodservice Management
Department of Applied Human
Ecology
106 Johnston Hall
Bowling Green, Ohio
43403-0254

Kent State University
Hospitality Foodservice
Management
School of Family and
Consumer Studies
103 Nixson Hall
Kent, Ohio 44242

Ohio State University
Hospitality Management
College of Human Ecology
1787 Neil Avenue, 265
Campbell Hall
Columbus, Ohio 43210

Tiffin University
Hotel and Restaurant
Management
155 Miami Street
Tiffin, Ohio 44883

Oklahoma

Langston University
Hospitality Management
Division of Business
P.O. Box 339
Langston, Oklahoma 73050

Oklahoma State University
School of Hotel and Restaurant
Administration
College of Home Economics
210 Home Economics West
Stillwater, Oklahoma 74078

Oregon

Oregon State University
Hotel, Restaurant and Tourism
Management
College of Business
Bexell Hall, Room 201
Corvallis, Oregon 97331-2603

Pennsylvania

Cheyney University
Hotel, Restaurant and
Institutional Management
Cheyney, Pennsylvania 19319

Drexel University
  Hotel, Restaurant, and
    Institutional Management
  College of Design Arts
  33rd and Market Street
  Philadelphia, Pennsylvania
    19104

East Stroudsburg University
  Hospitality Management
  School of Professional Studies
  Hospitality Management Center
  East Stroudsburg,
    Pennsylvania 18301

Indiana University of
    Pennsylvania
  Hotel, Restaurant, and
    Institutional Management
  Human Ecology and Health
    Science
  Ackerman Hall
  Indiana, Pennsylvania 15701

Marywood College
  Hotel and Restaurant
    Management
  2300 Adams Avenue
  Scranton, Pennsylvania
    18509-1598

Mercyhurst College
  Hotel, Restaurant and
    Institutional Management
  Glenwood Hills
  Erie, Pennsylvania 16546

Pennsylvania State University
  M School of Hotel, Restaurant
    and Institutional
    Management
  118 Henderson Building
  University Park, Pennsylvania
    16802

Widener University
  School of Hotel and Restaurant
    Management
  17th Street
  Chester, Pennsylvania 19013

Rhode Island

Johnson & Wales University
  M Hospitality Department
  Business Division
  8 Abbott Park Place
  Providence, Rhode Island
    02903

South Carolina

University of South Carolina
  Hotel, Restaurant and Tourism
    Administration
  College of Applied
    Professional Sciences
  Columbia, South Carolina
    29208

Winthrop College
  Food Systems Management
  Department of Human
    Nutrition
  103 Crawford Health Center
  Rock Hill, South Carolina
    29733

South Dakota

Black Hills State University
  Travel Industry Management
  College of Business and Public
    Affairs
  1200 University
  Spearfish, South Dakota 57783

Huron University
  Hotel and Restaurant
    Management
  Business Department
  8th and Ohio
  Huron, South Dakota 57350

South Dakota State University
  Restaurant Management
  College of Home Economics
  P.O. Box 2275A SDSU
  Brookings, South Dakota
    57007

Tennessee

Belmont College
  Hospitality Business
  1900 Belmont Boulevard
  Nashville, Tennessee 37212

Tennessee State University
  Foodservice Management
  Department of Home
    Economics
  3500 John Merritt Boulevard
  Nashville, Tennessee 37209

University of Tennessee
  Hotel and Restaurant
    Administration
  College of Human Ecology
  229 Jessie Harris Building
  Knoxville, Tennessee
    37996-1900

Texas

Huston-Tillotson College
  Hotel and Restaurant
    Management
  1820 East 8th Street
  Austin, Texas 78702

Lamar University
  Restaurant and Institutional
    Food Management
  School of Education
  P.O. Box 10035
  Beaumont, Texas 77710

Texas Tech University
  M Restaurant, Hotel and
    Institutional Management
  College of Home Economics
  Box 4170
  Lubbock, Texas 79409

Texas Woman's University
  Institution Administration
  Department of Nutrition and
    Food Sciences
  P.O. Box 24134, TWU Station
  Denton, Texas 76204

University of Houston
  M College of Hotel and
    Restaurant Management
  Houston, Texas 77204-3902

University of North Texas
  M Hotel and Restaurant
    Management
  School of Human Resource
    Management
  P.O. Box 5248
  Denton, Texas 76203

Wiley College
  Hotel and Restaurant
    Management
  711 Wiley Avenue
  Marshall, Texas 75670

Vermont

Johnson State College
  Hotel, Hospitality Management
  Johnson, Vermont 05656

Vermont College of Norwich
    University
  Hotel Administration Program
  Montpelier, Vermont 05602

Virginia

Liberty University
  Foodservice Management
  Department of Human Ecology
  Box 20000
  Lynchburg, Virginia
    24506-8001

James Madison University
  Hotel, Restaurant Management
  College of Business
  102 Harrison Hall
  Harrisonburg, Virginia 22807

Norfolk State University
  Hotel, Restaurant and
    Institutional Management
  School of Consumer Services
    and Family Studies
  2401 Corprew Avenue
  Norfolk, Virginia 23504

Radford University
  Restaurant Management
  Department of Health Service
  Norwood Street, Box 5826
  Radford, Virginia 24142

Virginia Polytechnic Institute and
    State University
  M Hotel, Restaurant and
    Institutional Management
  College of Human Resources
  18 Hillcrest Hall
  Blacksburg, Virginia
    24061-0429

Virginia State University
  Hotel Restaurant Management
  College of Agriculture and
    Applied Sciences
  Box 427 HRM
  Petersburg, Virginia 23803

Washington

Washington State University
Hotel and Restaurant
Administration Program
College of Business and
Economics
Todd Hall 245D
Pullman, Washington
99164-4724

Washington State University
Hotel and Restaurant
Administration
College of Business and
Economics
1108 East Columbia
Seattle, Washington 98122

West Virginia

Concord College
Travel Industry Management
Box 67
Athens, West Virginia 24712

Shepherd College
Hotel, Motel, Restaurant
Management
Shepherdstown, West Virginia
25401

Wisconsin

University of
Wisconsin—Madison
Foodservice Administration
College of Agricultural and
Life Sciences
1605 Linden Drive
Madison, Wisconsin 53706

University of
Wisconsin—Stevens Point
Food Systems Administration
School of Home Economics
CPS Building
Stevens Point, Wisconsin
54481

University of Wisconsin—Stout
Hotel and Restaurant
Management
School of Home Economics
Home Economics Building,
Room 220
Menomonie, Wisconsin 54751

Junior/Community Colleges

Alabama

Bessemer State Technical College
Institutional Foodservice
P.O. Box 308
Bessemer, Alabama 35021

Carver State Technical College
Commercial Foodservice
414 Stanton Street
Mobile, Alabama 36617

Community College of the Air
    Force
Foodservice and Lodging
CCAF/AYS Building 836
Maxwell AFB, Alabama 36112

Jefferson State Community
    College
Foodservice Management and
    Technology
2601 Carson Road
Birmingham, Alabama 35215

Lawson State Community College
Commercial Foods
3060 Wilson Road S.W.
Birmingham, Alabama 35221

Wallace State Community
    College
Commercial Foods
801 Main Street N.W.
Hanceville, Alabama
    35077-9080

Alaska

Alaska Pacific University
Travel and Hospitality
    Management
4101 University Drive
Anchorage, Alaska 99508

Alaska Vocational Technical
    Center
FoodService Technology
P.O. Box 889
Seward, Alaska 99664

University of Alaska
Foodservice Technology
3211 Providence Avenue
Anchorage, Alaska 99508

Arizona

Phoenix College
Foodservice Administration
1202 West Thomas Road
Phoenix, Arizona 85013

Pima County Community
    College District
Hospitality Department
P.O. Box 5027
1255 North Stone Avenue
Tucson, Arizona 85703-0027

Scottsdale Community College
Hospitality Management/
    Culinary Arts
9000 East Chaparral Road
Scottsdale, Arizona 85256

Arkansas

Quapaw Vocational Technical
    School
Foodservice Program
201 Vo Tech Drive
Hot Springs, Arkansas 71913

Southern Arkansas University
    Tech
Hotel, Restaurant Department
SAU Tech Station
Camden, Arkansas 71701

California

American River College
Foodservice Management
4700 College Oak Drive
Sacramento, California 95841

Bakersfield College
Foodservice Program
1801 Panorama Drive
Bakersfield, California 93305

Cabrillo College
Foodservice Technology
6500 Soquel Drive
Aptos, California 95003

California Culinary Academy
Culinary Arts/Chef Training
625 Polk Street
San Francisco, California
94102

Chaffey College
Hotel and Foodservice
Management
5885 Haven Avenue
Rancho Cucamonga,
California 91701

City College of San Francisco
Hotel and Restaurant
Department
50 Phelan Avenue
San Francisco, California
94112

Columbia Community College
Hospitality Management
P.O. Box 1849
Columbia, California 95310

Contra Costa
Culinary Arts
2600 Mission Bell Drive
San Pablo, California 93806

Cypress College
Culinary Arts and Hospitality
Management
9200 Valley View Boulevard
Cypress, California 90630

Diablo Valley College
Hotel, Restaurant Management
321 Golf Club Road
Pleasant Hill, California 94523

El Camino College
Foodservice Management
16007 Crenshaw
Torrance, California 90506

Glendale Community College
Foodservice Management
1500 North Verdugo Road
Glendale, California 91208

Grossmont College
Foodservice Management
8800 Grossmont College Drive
El Cajon, California 92020

Lake Tahoe Community College
Innkeeping and Restaurant
Operations
P.O. Box 14445
South Lake Tahoe, California
95702

Laney College
Culinary Arts Department
900 Fallon Street
Oakland, California 94607

Los Angeles Trade-Technical
  College
  Chef Training/Restaurant
  Management
  400 West Washington
  Boulevard
  Los Angeles, California 90015

Merced College
  Foodservice Program
  3600 M Street
  Merced, California 95348

Mission College
  Hospitality Management
  3000 Mission College
  Boulevard
  Santa Clara, California 95054

Modesto Junior College
  Foodservice Program
  435 College Avenue
  Modesto, California 95350

Orange Coast College
  Hotel Restaurant
  Management/Culinary Arts
  2701 Fairview Road
  Costa Mesa, California 92626

Oxnard College
  Hotel and Restaurant
  Management
  4000 South Rose Avenue
  Oxnard, California 93033

Pasadena City College
  Foodservice Program
  1570 East Colorado Boulevard
  Pasadena, California 91106

San Diego Mesa College
  Foodservice Occupations/
  Hotel, Motel Management
  7250 Mesa College Drive
  San Diego, California 92111

San Joaquin Delta College
  Foodservice Management/
  Foodservice Industry
  5151 Pacific Avenue
  Stockton, California 95207

Santa Barbara City College
  Hotel, Restaurant
  Management/Culinary Arts
  721 Cliff Drive
  Santa Barbara, California
  93109

Shasta College
  Culinary Arts
  1065 North Old Oregon Trail
  Redding, California 96001

Skyline College
  Hotel and Restaurant
  Operations
  3300 College Drive
  San Bruno, California 94066

Victor Valley College
  Restaurant Management
  18422 Bear Valley Road
  Victorville, California 92392

Yuba Community College
  Foodservice Management
  2088 North Beale Road
  Marysville, California 95901

Colorado

Career Development Center
Restaurant Careers
1200 South Sunset
Longmont, Colorado 80501

Colorado Mountain College
Resort Management
Box 5288
Steamboat Springs, Colorado
80477

Denver Institute of Technology
Hospitality Service
Management
7350 North Broadway
Denver, Colorado 80221

Emily Griffith Opportunity
School
Food Production Management
and Service
1250 Welton Street
Denver, Colorado 80204

Mesa State College
Travel, Recreation and
Hospitality Management
Box 2647, 1175 Texas Avenue
Grand Junction, Colorado
81502

T.H. Pickens Technical Center
Restaurant Arts
500 Buckley Road
Aurora, Colorado 80011

Pikes Peak Community College
Culinary Arts/Food
Management
5675 South Academy
Boulevard
Colorado Springs, Colorado
80906

Pueblo Community College
Foodservice Program
900 West Orman Avenue
Pueblo, Colorado 81004

Warren Occupational Technical
Center
Restaurant Arts
13300 West Ellsworth Avenue
Golden, Colorado 80401

Connecticut

Briarwood College
Hotel, Restaurant Management
2279 Mount Vernon Road
Southington, Connecticut
06489

Manchester Community College
Hotel, Foodservice
Management
60 Bidwell Street
Manchester, Connecticut 06040

Mattatuck Community College
Hospitality Management
750 Chase Parkway
Waterbury, Connecticut 06708

University of New Haven
Hotel, Restaurant, Tourism
Administration
300 Orange Avenue
West Haven, Connecticut
06516

Delaware

Delaware Tech—Southern
Campus
Hospitality Management
P.O. Box 610
Georgetown, Delaware 19947

District of Columbia

Culinary School of Washington,
Ltd.
Chef Programs
1634 I Street, N.W.
Washington, D.C. 20006

Florida

Atlantic Vocational Technical
Center
Culinary Arts
4700 N.W. Coconut Creek
Parkway
Coconut Creek, Florida 33066

Brevard Community College
Hospitality Management
1519 Clearlake Road
Cocoa, Florida 32922

Broward Community College
Restaurant Management
3501 S.W. Davie Road
Davie, Florida 33314

Daytona Beach Community
College
Hospitality Management
1200 Volusia Avenue
Daytona Beach, Florida 32115

Florida Community College at
Jacksonville
Culinary Arts/Hospitality
Management
3939 Roosevelt Boulevard
Jacksonville, Florida 32205

Florida Keys Community College
Hospitality Management
5901 West Junior College Road
Key West, Florida 33040

Gulf Coast Community College
Hotel, Motel and Restaurant
Management
5230 West Highway 98
Panama City, Florida 32401

Hillsborough Community College
Hotel and Resort
Management/Chef
Apprentice Training
P.O. Box 30030
Tampa, Florida 33630

Indian River Community College
Restaurant Management
3209 Virginia Avenue
Fort Pierce, Florida 33454

Manatee Community College
Hospitality Management
5840 26th Street West
Bradenton, Florida 33507

Miami-Dade Community College
Hospitality Management
300 N.E. 2nd Avenue
Miami, Florida 33132

Mid Florida Technical Institute
Commercial Cooking and
Culinary Arts
2900 West Oakridge Road
Orlando, Florida 32809

North Technical Educational
Center
Commercial Foods/Culinary
Arts
7071 Garden Road
Riviera Beach, Florida 33404

Okaloosa-Walton Community
College
Commercial Foods
100 College Boulevard
Niceville, Florida 32578

Palm Beach Community College
Hospitality Management
4200 South Congress Avenue
Lake Worth, Florida 33461

Pensacola Junior College
Hospitality Management
1000 College Boulevard
Pensacola, Florida 32504-8998

PETEC—Clearwater Campus
Culinary Arts
6100 154th Avenue North
Clearwater, Florida 34620

Sarasota County Vocational
Technical Center
Commercial Cooking/Culinary
Arts
4748 Beneva Road
Sarasota, Florida 34233

Seminole Community College
Culinary Arts/Restaurant
Management
100 Weldon Boulevard
Sanford, Florida 32773

St. Augustine Technical Center
Commercial Foods/Culinary
Arts
2960 Collins Avenue
St. Augustine, Florida 32084

Valencia Community College
Hospitality Management
P.O. Box 3028
Orlando, Florida 32802

Washington-Holmes Area
Vocational Technical Center
Commercial Foods and
Culinary Arts
209 Hoyt Street
Chipley, Florida 32428

Webber College
Hotel and Restaurant
Management
1201 Alternate Highway 27
South
Babson Park, Florida 33827

Georgia

Albany Technical Institute
  Culinary Arts
  1021 Lowe Road
  Albany, Georgia 30310

Atlanta Technical Institute
  Culinary Arts
  1560 Stewart Avenue, S.W.
  Atlanta, Georgia 30310

Augusta Technical Institute
  Culinary Arts
  3116 Deans Bridge Road
  Augusta, Georgia 30906

Ben Hill–Irwin Technical Institute
  Culinary Arts
  P.O. Box 1069
  Fitzgerald, Georgia 31750

Gainesville College
  Hotel, Restaurant and Travel
  P.O. Box 1358
  Gainesville, Georgia 30503

Gwinnett Technical Institute
  Hotel, Restaurant, Travel
    Management
  P.O. Box 1505
  1250 Atkinson Road
  Lawrenceville, Georgia 30246

Macon Technical Institute
  Culinary Arts
  3300 Macon Tech Drive
  Macon, Georgia 31206

Middle Georgia Technical
    Institute
  Culinary Arts
  1311 Corder Road
  Warner Robins, Georgia 31088

Savannah Technical Institute
  Culinary Arts
  5717 White Bluff Road
  Savannah, Georgia 31499

Hawaii

Hawaii Community College
  Foodservice Program
  1175 Manono Street
  Hilo, Hawaii 96720-4091

Kapiolani Community College
  Foodservice and Hospitality
    Education
  4303 Diamond Head Road
  Honolulu, Hawaii 96816

Leeward Community College
  Foodservice Program
  96-045 Ala Ike
  Pearl City, Hawaii 96782

Maui Community College
  Foodservice Program
  310 Kaahumanu Avenue
  Kahului, Hawaii 96732

Idaho

Boise State University
  Culinary Arts
  1910 University Drive
  Boise, Idaho 83725

College of Southern Idaho
  Hotel, Restaurant Management
  P.O. Box 1238
  Twin Falls, Idaho 83303

Ricks College
  Restaurant and Catering
    Management
  Clark Building
  Rexburg, Idaho 83440

Illinois

Chicago Hospitality Institute
  Foodservice Administration,
    Hotel, Motel Management
  226 West Jackson Boulevard
  Chicago, Illinois 60606

College of DuPage
  Hospitality Administration
  22nd Street and Lambert Road
  Glen Ellyn, Illinois 60137

College of Lake County
  Foodservice Management/
    Culinary Arts
  19351 West Washington
  Grayslake, Illinois 60030

The Cooking and Hospitality
    Institute of Chicago
  Professional
    Cooking/Hospitality
    Management
  361 West Chestnut
  Chicago, Illinois 60610

Elgin Community College
  Culinary Arts/Hospitality
    Management
  1700 Spartan Drive
  Elgin, Illinois 60123

William Rainey Harper
    Community College
  Foodservice
    Management/Culinary Arts
  1200 West Algonquin Road
  Palatine, Illinois 60067

Joliet Junior College
  Culinary Arts/Hotel Restaurant
    Management
  1216 Houbolt
  Joliet, Illinois 60436

Kendall College
  Culinary Arts/Hospitality
    Management
  2408 Orrington Avenue
  Evanston, Illinois 60201

Kennedy-King College
  Food Management
  6800 South Wentworth Avenue
  Chicago, Illinois 60621

Lewis & Clark Community
    College
  Hospitality Industry Programs
  5800 Godfrey Road
  Godfrey, Illinois 62035

Lexington Institute
  Hospitality Careers
  10840 South Western Avenue
  Chicago, Illinois 60643

Lincoln Trail College
  Foodservice Technology
  Route #3
  Robinson, Illinois 62454

Moraine Valley Community
College
Restaurant Management
10900 South 88th Avenue
Palos Hills, Illinois 60465

Oakton Community College
Hotel Management
1600 East Golf Road
Des Plaines, Illinois 60016

Parkland College
Hospitality Industries
2400 West Bradley
Champaign, Illinois 61821

Sauk Valley Community College
Foodservice Program
173 Illinois Route 2
Dixon, Illinois 61021

Triton College
Hospitality Institute
2000 Fifth Avenue
River Grove, Illinois 60171

Washburne Trade School
Chefs Training
3233 West 31st Street
Chicago, Illinois 60623

Indiana

Ball State University
Foodservice Management
Practical Arts Building
Muncie, Indiana 47306

Indiana-Purdue University at
Indianapolis
Restaurant, Hotel and
Institutional Management
799 West Michigan Street
Indianapolis, Indiana 46202

Indiana Vocational Technical
College
Culinary Arts
5727 Sohl Avenue
Hammond, Indiana 46320

Indiana Vocational Technical
College
Hotel and Restaurant
Administration/Culinary
Arts
P.O. Box 1763
Indianapolis, Indiana 46206

Purdue University
Restaurant, Hotel and
Institutional Management
106 Stone Hall
West Lafayette, Indiana 47907

Vincennes University
Hospitality Management/
Culinary Arts
1002 North 1st Street
Vincennes, Indiana 47591

Iowa

Des Moines Area Community
College
Hospitality Careers
2006 Ankeny Boulevard
Ankeny, Iowa 50021

Indian Hills Community College
  Foodservice/Cooking
  525 Grandview
  Ottumwa, Iowa 52501

Iowa Lakes Community College
  Hotel, Motel and Restaurant
    Management
  3200 College Drive
  Emmetsburg, Iowa 50536

Iowa Western Community College
  Culinary Arts
  2700 College Road, Box 4-C
  Council Bluffs, Iowa 51502

Kirkwood Community College
  Restaurant Management/
    Culinary Arts
  6301 Kirkwood Boulevard
    S.W.
  Cedar Rapids, Iowa 52406

Kansas

Central College
  Foodservice Management
  1200 South Main
  McPherson, Kansas 67460

Cloud County Community
    College
  Hospitality Management
  2221 Campus Drive
  Concordia, Kansas 66901

Flint Hills Area Vocational
    Technical School
  Culinary Arts
  3301 West 18th Avenue
  Emporia, Kansas 66801

Hesston College
  Hotel, Restaurant, Institutional
    Management
  Box 3000
  Hesston, Kansas 67072

Johnson County Community
    College
  Hospitality Program
  12345 College
  Overland Park, Kansas 66210

Kansas City Area Vocational
    Technical School
  Commercial Foodservice
  2220 North 59th Street
  Kansas City, Kansas 66104

Kaw Area Vocational Technical
    Institute
  Foodservice Program
  5724 Huntoon
  Topeka, Kansas 66604

Northeast Kansas Area
    Vocational Technical School
  Quantity Foods
  1501 West Riley
  Atchison, Kansas 66002

Salina Area Vocational Technical
    School
  Foodservice Management
  2562 Scanlan
  Salina, Kansas 67401

Wichita Area Vocational
    Technical School
  Food Service
    Mid-Management and
    Commercial Cooking
  324 North Emporia
  Wichita, Kansas 67202

Kentucky

Daviess County State Vocational
  Technical School
  Commercial Foods
  1901 S.E. Parkway
  Owensboro, Kentucky 42303

Elizabethtown State Technical
  School
  Commercial Foods
  505 University Drive
  Elizabethtown, Kentucky
  42701

Jefferson Community College
  Culinary Arts
  109 East Broadway
  Louisville, Kentucky 40202

Sullivan College
  Culinary Arts/Hotel,
  Restaurant Management
  3101 Bardstown Road
  Louisville, Kentucky 40205

West Kentucky State Vocational
  Technical School
  Culinary Arts
  P.O. Box 7408
  Paducah, Kentucky 42001

Louisiana

Baton Rouge Vocational
  Technical Institute
  Culinary Occupations
  3250 North Acadian
  Throughway
  Baton Rouge, Louisiana 70805

Sidney N. Collier Vocational
  Technical School
  Culinary Occupations
  3727 Louisa Street
  New Orleans, Louisiana 70126

Delgado Community College
  Culinary Apprenticeship
  Program
  615 City Park Avenue
  New Orleans, Louisiana 70119

New Orleans Regional Vocational
  Technical Institute
  Culinary Arts
  980 Navarre Avenue
  New Orleans, Louisiana 70124

Nicholls State University
  Food Management
  P.O. Box 2014
  Thibodaux, Louisiana 70310

Southern University
  Hotel, Motel, Restaurant
  Management
  3050 Martin Luther King Drive
  Shreveport, Louisiana 71107

Maine

Eastern Maine Technical College
  Food Management/Food
  Technology
  354 Hogan Road
  Bangor, Maine 04401

Southern Maine Technical
College
Hotel, Motel and Restaurant
Management and Culinary
Arts
Fort Road
South Portland, Maine 04106

Washington County Technical
College
Foodservice Program
River Road
Calais, Maine 04619

Maryland

Allegany Community College
Foodservice Management
Willow Brook Road
Cumberland, Maryland 21502

Anne Arundel Community
College
Hotel, Restaurant Management
101 College Parkway C-205
Arnold, Maryland 21012

Baltimore's International
Culinary College
Culinary Arts/Restaurant
Management
19-21 South Gay Street
Baltimore, Maryland 21202

Essex Community College
Hotel, Motel, Restaurant, Club
Management
7201 Rossville Boulevard
Baltimore County, Maryland
21237

Hagerstown Junior College
Hospitality Industry
751 Robinwood Drive
Hagerstown, Maryland 21740

Howard Community College
Culinary Apprenticeship/
Restaurant Management
10920 Route 108
Ellicott City, Maryland 21043

Montgomery College
Hospitality Management
51 Mannakee Street
Rockville, Maryland 20850

Prince George's Community
College
Hospitality Services
Management
301 Largo Road
Largo, Maryland 20772

Wor-Wic Tech Community
College
Hotel, Motel, Restaurant
Management
Route 3, Box 79
Berlin, Maryland 21811

Massachusetts

Bay Path College
Hotel and Hospitality
Management
588 Long Meadow Street
Long Meadow, Massachusetts
01106

Becker Junior College
  Resort, Hotel, Restaurant
    Management
  3 Paxton Street
  Leicester, Massachusetts 01524

Berkshire Community College
  Hotel and Restaurant
    Management
  West Street
  Pittsfield, Massachusetts 01201

Bunker Hill Community College
  Hotel, Restaurant, Travel
    Management
  New Rutherford Avenue
  Boston, Massachusetts 02129

Cape Cod Community College
  Hotel, Restaurant Management
  Route 132
  West Barnstable,
    Massachusetts 02668

Endicott College
  Hotel, Restaurant, and Travel
    Administration
  376 Hale Street
  Beverly, Massachusetts 01915

Katherine Gibbs School
  Hotel, Restaurant Management
  5 Arlington Street
  Boston, Massachusetts 02116

Holyoke Community College
  Hospitality Management
  303 Homestead Avenue
  Holyoke, Massachusetts 01040

Massachusetts Bay Community
    College
  Hospitality Management
  Fay Road
  Framingham, Massachusetts
    01701

Massasoit Community College
  Culinary Arts
  1 Massasoit Boulevard
  Brockton, Massachusetts
    02402

Mount Ida College
  Hotel, Institution Management
  777 Dedham Street
  Newton Center, Massachusetts
    02159

Newbury College
  Hospitality Management and
    Culinary Arts
  129 Fisher Avenue
  Brookline, Massachusetts
    02146

Northeastern University
  Hotel and Restaurant
    Management/Culinary Arts
  270 Ryder Building
  Boston, Massachusetts 02115

Quincy Junior College
  Hospitality Management
  34 Coddington Street
  Quincy, Massachusetts 02169

Quinsigamond Community
College
Hotel and Restaurant
Management
670 West Boylston Street
Worcester, Massachusetts
01606-2092

Michigan

Davenport College of Business
Restaurant and Lodging
Management
415 East Fulton
Grand Rapids, Michigan 49507

Henry Ford Community College
Hospitality Studies
5101 Evergreen
Dearborn, Michigan 48128

Gogebic Community College
Foodservice and Hospitality
Management
E4946 Jackson Road
Ironwood, Michigan 49938

Grand Rapids Junior College
Culinary Arts, Food and
Beverage Management
143 Bostwick, N.E.
Grand Rapids, Michigan 49503

Kalamazoo Valley Community
College
Foodservice Management
6767 West O Avenue
Kalamazoo, Michigan 49009

Lake Michigan College
Food Management
2755 East Napier
Benton Harbor, Michigan
49022

Lansing Community College
Hospitality Systems
419 North Capitol Avenue
Lansing, Michigan 48901

Macomb Community College
Culinary Arts/Professional
Foodservice
44575 Garfield Road
Mount Clemens, Michigan
48044

Mott Community College
Foodservice Management/
Culinary Arts
1401 East Court Street
Flint, Michigan 48503

Muskegon Community College
Foodservice, Lodging and
Travel Management
221 South Quarterline Road
Muskegon, Michigan 49442

Northern Michigan University
Restaurant Foods
Jacobetti Center, Route 550
Marquette, Michigan 49855

Northwestern Michigan College
Foodservice and Hospitality
Management
1701 East Front Street
Traverse City, Michigan 49684

Northwood Institute
  Hotel, Restaurant Management
  3225 Cook Road
  Midland, Michigan 48640

Oakland Community College
  Hospitality Management/
    Culinary Arts
  27055 Orchard Lake Road
  Farmington Hills, Michigan
    48018

People's Community Civic
    League
  Culinary Arts
  5961 14th Street
  Detroit, Michigan 48208

Schoolcraft College
  Culinary Arts and Culinary
    Management
  18600 Haggerty Road
  Livonia, Michigan 48152

Siena Heights College
  Hotel, Restaurant and
    Institutional Management
  1247 East Siena Heights Drive
  Adrian, Michigan 49221

State Technical Institute
  Culinary Arts
  Alber Drive
  Plainwell, Michigan 49080

Sumomi College
  Hotel, Restaurant Management
  601 Quincy Street
  Hancock, Michigan 49930

Washtenaw Community College
  Culinary Arts/Hospitality
    Management
  4800 East Huron River Drive
  Ann Arbor, Michigan 48106

Wayne County Community
    College
  Culinary Arts
  8551 Greenfield
  Detroit, Michigan 48228

West Shore Community College
  Foodservice Management
  3000 Stiles Road
  Scottville, Michigan 49454

Minnesota

Alexandria Technical College
  Hotel, Restaurant Management
  1601 Jefferson
  Alexandria, Minnesota 56308

Dakota County Technical College
  Foodservice Management
  1300 East 145th Street
  Rosemont, Minnesota 55068

Detroit Lakes Technical College
  Chef Training and Commercial
    Cooking
  East Highway 34
  Detroit Lakes, Minnesota
    56501

Duluth Technical College
  Culinary Arts/Restaurant
    Management
  2101 Trinity Road
  Duluth, Minnesota 55811

Hennepin Technical College
Cook/Chef
9000 Brooklyn Boulevard
Brooklyn Park, Minnesota
55445

Mankato Technical College
Culinary Arts
1920 Lee Boulevard
North Mankato, Minnesota
56001

Minneapolis Technical College
Culinary Arts/Hospitality
Management
1415 Hennepin Avenue
Minneapolis, Minnesota 55403

Moorhead Technical College
Chef Training
1900 28th Avenue South
Moorhead, Minnesota 56560

Normandale Community College
Hospitality Management
9700 France Avenue South
Bloomington, Minnesota
55431

Northeast Metro Technical
College
Quantity Foods/Chef Training
3300 Century Avenue North
White Bear Lake, Minnesota
55110

Southwestern Technical College
Foodservice Department
North Hiawatha Avenue
Pipestone, Minnesota 56164

St. Paul Technical College
Restaurant and Hotel Cookery
235 Marshall Avenue
St. Paul, Minnesota 55102

University of Minnesota,
Crookston
Hospitality Department
Highways 2 and 75 North
Crookston, Minnesota 56716

Mississippi

Hinds Community College
Hotel, Restaurant Management
Technology
3925 Sunset Drive
Jackson, Mississippi 39213

Meridian Community College
Restaurant and Hotel
Management
5500 Highway 19 North
Meridian, Mississippi 39307

Mississippi Gulf Coast
Community College
Motel, Restaurant Technology
2226 Switzer Road
Gulfport, Mississippi 39507

Northeast Mississippi
Community College
Hotel, Restaurant Management
Technology
Cunningham Boulevard
Booneville, Mississippi 38829

Missouri

Crowder College
  Hospitality Management
  601 Laclede
  Neosho, Missouri 64850

Jefferson College
  Hotel, Restaurant Management
  P.O. Box 1000
  Hillsboro, Missouri 63050

Penn Valley Community College
  Lodging and Foodservice
    Management/Culinary Arts
  3201 Southwest Trafficway
  Kansas City, Missouri 64152

St. Louis Community College at
    Forest Park
  Hospitality Restaurant
    Management
  5600 Oakland Avenue
  St. Louis, Missouri 63110

Montana

Missoula Vocational Technical
    Center
  Commercial Food Production
  909 South Avenue West
  Missoula, Montana 59801

Nebraska

Central Community College
  Hotel, Motel, Restaurant
    Management
  P.O. Box 1024
  Hastings, Nebraska 68901

Metropolitan Community College
  Foodservice Technology
  P.O. Box 3777, 30th and Fort
    Streets
  Omaha, Nebraska 68103

Nebraska College of Business
  Hotel, Restaurant Management
  3636 California Street
  Omaha, Nebraska 68131

Southeast Community College
  Foodservice Program
  8800 O Street
  Lincoln, Nebraska 68520

Nevada

Clark County Community
    College
  Hotel, Restaurant, Casino
    Management and Culinary
    Arts
  3200 East Cheyenne Avenue
  North Las Vegas, Nevada
    89030

Truckee Meadows Community
    College
  Foodservice Techniques
  7000 Dandini Boulevard
  Reno, Nevada 89512

New Hampshire

Hesser College
  Hotel, Restaurant Management
  25 Fowell Street
  Manchester, New Hampshire
    03101

New Hampshire College
  Hotel, Restaurant Management
    and Culinary Arts
  2500 North River Road
  Manchester, New Hampshire
    03104

New Hampshire Technical
    College
  Culinary Arts
  2020 Riverside Drive
  Berlin, New Hampshire 03570

University of New Hampshire
  Foodservice Management
  Barton Hall, Room 105
  Durham, New Hampshire
    03824

New Jersey

Atlantic Community College
  Culinary Arts/Hospitality
    Management
  Black Horse Pike
  Mays Landing, New Jersey
    08330

Bergen Community College
  Hotel, Restaurant Management
  400 Paramus Road
  Paramus, New Jersey 07652

Brookdale Community College
  Foodservice Management
  Newman Springs Road
  Lincroft, New Jersey 07738

Burlington County College
  Hospitality Management
  Route 530
  Pemberton, New Jersey 08068

Camden County College
  Food Management
  Little Gloucester Road, Box
    200
  Blackwood, New Jersey 08012

Cape May County Vocational
    Technical Schools
  Foods Production/Culinary
    Arts
  Crest Haven Road
  Cape May Court House, New
    Jersey 08210

County College of Morris
  Hotel, Restaurant Management
  Center Grove Road
  Randolph, New Jersey 07869

Hudson County Community
    College
  Culinary Arts
  161 Newkirk Street
  Jersey City, New Jersey 07306

Mercer County Community
    College
  Hotel, Restaurant and
    Institution Management
  1200 Old Trenton Road
  Trenton, New Jersey 08690

Middlesex County College
  Hotel, Restaurant and
    Institution Management
  155 Mill Road, Box 3050
  Edison, New Jersey 08818

Ocean County College
  Foodservice Management
  College Drive
  Toms River, New Jersey 08753

Union County Vocational
Technical School
Foodservice Program
1776 Raritan Road
Scotch Plains, New Jersey
07076

New Mexico

Albuquerque Technical
Vocational Institute
Culinary Arts
525 Buena Vista S.E.
Albuquerque, New Mexico
87059

New York

Adirondack Community College
Foodservice Program
Bay Road
Queensbury, New York 12804

Community College of the
Finger Lakes
Hotel and Resort Management
Lincoln Hill
Canandaigua, New York 14424

Culinary Institute of America
Culinary Arts
Route 9
Hyde Park, New York 12538

Erie Community College North
Foodservice Administration/
Restaurant Management
Main and Youngs Road
Buffalo, New York 14221

Fulton Montgomery Community
College
Foodservice Administration
Route 67
Johnstown, New York 12866

Genesee Community College
Hospitality Management
One College Road
Batavia, New York 14020

Jefferson Community College
Hospitality and Tourism
Management
Outer Coffeen Street
Watertown, New York 13601

LaGuardia Community College
Commercial Foodservice
Management
31 - 10 Thomson Avenue
Long Island City, New York
11101

Mohawk Valley Community
College
Foodservice Program
Upper Floyd Avenue
Rome, New York 13440

Monroe Community College
Food, Hotel and Tourism
Management
1000 East Henrietta Road
Rochester, New York 14623

Nassau Community College
Hotel and Restaurant
Management
Building K
Garden City, New York 11530

New York City Technical College
  Hotel and Restaurant
    Management
  300 Jay Street
  Brooklyn, New York 11201

New York Food and Hotel
  Management School
  Hospitality Management
  154 West 14th Street
  New York, New York 10011

New York Institute of Technology
  Culinary Arts
  Carleton Avenue
  Central Islip, New York
    11722-4597

New York Restaurant School
  Culinary Arts/Restaurant
    Management
  27 West 34th Street
  New York, New York 10001

Niagara County Community
    College
  Professional Chef
  3111 Saunders Settlement Road
  Sanborn, New York 14132

Onondaga Community College
  Foodservice Administration/
    Hotel Technology
  Route 173
  Syracuse, New York 13215

Rockland Community College
  Foodservice Management
  145 College Road
  Suffern, New York 10901

Schenectady County Community
    College
  Hotel, Culinary Arts and
    Tourism
  78 Washington Avenue
  Schenectady, New York 12305

Paul Smith's College
  Hotel Restaurant
    Management/Culinary Arts
  Paul Smith's, New York 12970

State University of New York at
    Alfred
  Foodservice Department
  South Brooklyn Avenue
  Wellsville, New York 14895

State University of New York at
    Canton
  Hotel and Restaurant
    Management
  Cornell Drive
  Canton, New York 13617

State University of New York at
    Cobleskill
  Foodservice and Hospitality
    Administration
  Champlin Hall
  Cobleskill, New York 12043

State University of New York at
    Delhi
  Hospitality Management
  Alumni Hall
  Delhi, New York 13753

State University of New York at
    Farmingdale
  Restaurant Management
  Melville Road, Thompson Hall
  Farmingdale, New York 11735

State University of New York at
  Morrisville
Food Administration,
  Restaurant Management
Bailey Annex
Morrisville, New York 13408

Suffolk County Community
  College
Hotel, Restaurant Management
Speonk-Riverhead Road
Riverhead, New York 11901

Sullivan County Community
  College
Hospitality Program
Leroy Road
Loch Sheldrake, New York
  12759

Tompkins Cortland Community
  College
Restaurant Management
170 North Street
Dryden, New York 13053

Villa Maria College of Buffalo
Foodservice Management
240 Pine Ridge Road
Buffalo, New York 14225

Westchester Community College
Restaurant Management
75 Grasslands Road
Valhalla, New York 10595

North Carolina

Alamance Community College
Foodservice Management
P.O. Box 623
Haw River, North Carolina
  27258

Asheville-Buncombe Technical
  College
Hospitality Management
  Administration/Culinary
  Arts
340 Victoria Road
Asheville, North Carolina
  28804

Central Piedmont Community
  College
Hotel, Restaurant Management
P.O. Box 35009
Charlotte, North Carolina
  28235

Fayetteville Technical
  Community College
Foodservice Management
P.O. Box 35236
Fayetteville, North Carolina
  28303

Guilford Technical Community
  College
Culinary Arts Technology
P.O. Box 309
Jamestown, North Carolina
  27282

Lenoir Community College
Foodservice Management
P.O. Box 188
Kinston, North Carolina 28502

Southwestern Community
  College
Foodservice Management
275 Webster Road
Sylva, North Carolina 28779

Wake Technical Community
    College
  Hotel, Restaurant Management
    and Culinary Arts
  9101 Fayetteville Road
  Raleigh, North Carolina 27603

Wilkes Community College
  Hotel, Restaurant Management
  Drawer 120
  Wilkesboro, North Carolina
    28697

North Dakota

Bismarck State College
  Hotel, Restaurant Management
  1500 Edwards Avenue
  Bismarck, North Dakota 58501

North Dakota State College of
    Science
  Chef Training and
    Management Technology
  North 6th Street
  Wahpeton, North Dakota 58076

Ohio

Cincinnati Technical College
  Hotel, Restaurant
    Management/Chef Program
  3520 Central Parkway
  Cincinnati, Ohio 45223

Clermont College of the
    University of Cincinnati
  Hospitality Management
    Technology
  College Drive
  Batavia, Ohio 45103

Columbus State Community
    College
  Hospitality Management
  550 East Spring Street
  Columbus, Ohio 43215

Cuyahoga Community College
  Hospitality Management
  2900 Community College
    Avenue
  Cleveland, Ohio 44115

Hocking Technical College
  Hotel, Restaurant
    Management/Culinary Arts
  Hocking Parkway
  Nelsonville, Ohio 45764

Jefferson Technical College
  Foodservice Management
  4000 Sunset Boulevard
  Steubenville, Ohio 43952

Owens Technical College
  Hospitality Management
  P.O. Box 10,000, Oregon Road
  Toledo, Ohio 43699

Tiffin University
  Hotel and Restaurant
    Management
  155 Miami Street
  Tiffin, Ohio 44883

University of Akron
  Hotel, Motel, Restaurant
    Management/Culinary Arts
  200 East Exchange Street
  Akron, Ohio 44325

University of Toledo Community
and Technical College
Foodservice
Management/Culinary Arts
Scott Park Campus
Toledo, Ohio 43606

Youngstown State University
Hospitality Management
410 Wick Avenue
Youngstown, Ohio 44555

Oklahoma

Great Plains Area Vocational
Technical Center
Foodservice Management
4500 West Lee Boulevard
Lawton, Oklahoma 73505

Indian Meridian Vocational
Technical School
Commercial Food Production
1312 South Sangre Road
Stillwater, Oklahoma 74074

Oklahoma State University,
Technical Branch
Foodservice Management,
Culinary Arts
4th and Mission
Okmulgee, Oklahoma 74447

Pioneer Area Vocational
Technical School
Commercial Foods
2101 North Ash
Ponca City, Oklahoma 74601

Southern Oklahoma Area
Vocational Technical Center
Culinary Arts
Route #1, Box 14M
Ardmore, Oklahoma 73401

Oregon

Central Oregon Community
College
Hotel, Restaurant Management
2600 N.W. College Way
Bend, Oregon 97701

Chemeketa Community College
Hospitality Systems
P.O. Box 14007, 4000
Lancaster Drive N.E.
Salem, Oregon 97309

Lane Community College
Hospitality/Culinary Arts
4000 East 30th Avenue
Eugene, Oregon 97405

Linn-Benton Community College
Culinary Arts and Hospitality
Services
6500 S.W. Pacific Boulevard
Albany, Oregon 97321

Mt. Hood Community College
Hospitality and Tourism
26000 S.E. Stark
Gresham, Oregon 97030

Portland Community College
Hospitality Program
P.O. Box 19000
Portland, Oregon 97219-0990

Western Culinary Institute
Culinary Arts
1316 S.W. 13th Avenue
Portland, Oregon 97201

Pennsylvania

Bucks County Community
College
Hotel Restaurant Management,
Chef Apprenticeship
Swamp Road
Newton, Pennsylvania 18940

Butler County Community
College
Restaurant and Foodservice
Management
P.O. Box 1203, College Drive
Butler, Pennsylvania
16003-1203

Central Pennsylvania Business
School
Hotel, Motel Management
College Hill Road
Summerdale, Pennsylvania
17093-0309

Community College of
Allegheny County
Hospitality Management
595 Beatty Road
Monroeville, Pennsylvania
15146

Community College of
Philadelphia
Hotel, Restaurant and
Institutional
Management/Chef
Apprenticeship
1700 Spring Garden Street
Philadelphia, Pennsylvania
19130

Delaware County Community
College
Hotel, Restaurant Management
Route 252
Media, Pennsylvania 19063

Harcum Junior College
Hospitality, Tourism Program
Morris and Montgomery
Avenues
Bryn Mawr, Pennsylvania
19010

Harrisburg Area Community
College
Hotel, Restaurant and
Institutional Management
3300 Cameron Street
Harrisburg, Pennsylvania
17110

Keystone Junior College
Hospitality Management
Box 50
La Plume, Pennsylvania
18440-0220

Lehigh County Community
College
Hotel, Restaurant Management
2370 Main Street
Schnecksville, Pennsylvania
18078

Luzerne County Community
College
Hotel, Restaurant Management
Prospect Street and Middle
Road
Nanticoke, Pennsylvania 18634

Montgomery County Community
College
Hotel, Restaurant Management
340 Dekalb Pike
Bluebell, Pennsylvania 19422

Mount Aloysius Junior College
Hotel, Restaurant Management
William Penn Highway
Cresson, Pennsylvania 16630

Northampton Community College
Restaurant Management
3835 Green Pond Road
Bethlehem, Pennsylvania
18017

Peirce Junior College
Hospitality Management
1420 Pine Street
Philadelphia, Pennsylvania
19102

Pennsylvania College of
Technology
Food and Hospitality and
Culinary Arts
One College Avenue
Williamsport, Pennsylvania
17701

Pennsylvania State University,
Berks Campus
Hotel, Restaurant, and
Institutional Management
P.O. Box 7009
Reading, Pennsylvania
19610-6009

The Restaurant School
Restaurant Management/Chef
Training
2129 Walnut Street
Philadelphia, Pennsylvania
19103

Westmoreland County
Community College
Foodservice, Hotel, Motel
Management/Culinary Arts
College Station Road
Youngwood, Pennsylvania
15697-1895

Widener University
Hotel and Restaurant
Management
13th Street
Chester, Pennsylvania 19103

Rhode Island

Johnson & Wales University
  Hospitality Management/
  Culinary Arts
  8 Abbott Park Place
  Providence, Rhode Island
  02903

Rhode Island School of Design
  Culinary Arts Apprenticeship
  2 College Street
  Providence, Rhode Island
  02903

South Carolina

Anderson College
  Hotel, Restaurant and Tourism
  316 Boulevard
  Anderson, South Carolina
  29621

Greenville Technical College
  Foodservice Management
  P.O. Box 5616, Station B
  Greenville, South Carolina
  29606-5616

Horry Georgetown Technical
  College
  Hotel, Restaurant Management
  P.O. Box 1966, Route 501 East
  Conway, South Carolina 29526

Johnson & Wales University at
  Charleston
  Culinary Education,
  Hospitality Department
  701 East Bay Street
  Charleston, South Carolina
  29403

Technical College of the
  Lowcountry
  Hotel, Motel and Restaurant
  Management
  P.O. Box 1288
  100 S. Rabiut Road
  Beaufort, South Carolina
  29901

Trident Technical College
  Hospitality Department
  P.O. Box 10367, HT-P
  Charleston, South Carolina
  29411

South Dakota

Black Hills State University
  Travel Industry Management
  1200 University
  Spearfish, South Dakota 57783

Mitchell Vocational Technical
  Institute
  Chef Training
  821 North Capital
  Mitchell, South Dakota 57301

Tennessee

Knoxville State Area Vocational
  School
  Foodservice Program
  1100 Liberty Street
  Knoxville, Tennessee 37919

Memphis Culinary Academy
  Professional Culinary Arts
  1252 Peabody Avenue
  Memphis, Tennessee 38104

State Technical Institute at
    Memphis
Hotel, Restaurant Management
5983 Macon Cove
Memphis, Tennessee 38134

Texas

Central Texas College
    Foodservice Management
    P.O. Box 1800
    Killeen, Texas 76540

Del Mar College
    Restaurant Management
        Department
    Baldwin at Ayers
    Corpus Christi, Texas 78404

El Centro College
    Food and Hospitality Services
    Main at Lamar
    Dallas, Texas 75202

El Paso Community College
    Hospitality Travel Services
    P.O. Box 20500
    El Paso, Texas 79995

Galveston College
    Foodservice Management/
        Culinary Arts
    4015 Avenue Q
    Galveston, Texas 77550

Houston Community College
    Hotel, Restaurant
        Management/Culinary Arts
    1300 Holman
    Houston, Texas 77004

Lamar University
    Restaurant and Institutional
        Food Management
    P.O. Box 10035
    Beaumont, Texas 77710

Le Chef Culinary Arts School
    Culinary Arts
    6020 Dillard Circle
    Austin, Texas 78752

Northwood Institute
    Hotel and Restaurant
        Management
    Farm Road 1382
    Cedar Hill, Texas 75104-0058

San Jacinto College—North
        Campus
    Chef's Training
    5800 Uvaldo
    Houston, Texas 77049

South Plains College—Lubbock
    Food Industry Management
    1302 Main Street
    Lubbock, Texas 79401

St. Phillip's College
    Hospitality Operations
    2111 Nevada
    San Antonio, Texas 78203

Texas State Technical Institute
    Foodservice Technology
    3801 Campus Drive
    Waco, Texas 76705

Utah

Sevier Valley Tech
Foodservices/Cooking
800 West 200 South
Richfield, Utah 84701

Utah Valley Community College
Hotel, Motel, Restaurant
Management
800 West 1200 South
Orem, Utah 84058

Vermont

Champlain College
Hotel, Restaurant Management
P.O. Box 670
Burlington, Vermont 05402

New England Culinary Institute
Culinary Arts
250 Main Street
Montpelier, Vermont 05602

Virginia

Thomas Nelson Community
College
Hotel, Restaurant and
Institutional Management
P.O. Box 9407
Hampton, Virginia 23670

Northern Virginia Community
College
Hotel, Restaurant and
Institutional Management
8333 Little River Turnpike
Annandale, Virginia 22003

Tidewater Community College
Hotel, Restaurant and
Institutional Management
1700 College Crescent
Virginia Beach, Virginia
23456

Washington

Clark Community College
Culinary Arts/Restaurant
Management
1800 East McLoughlin
Boulevard
Vancouver, Washington 98663

Edmonds Community College
Culinary Arts
20000 68th Avenue West
Lynnwood, Washington 98036

Everett Community College
Food Technology
801 Wetmore Avenue
Everett, Washington 98201

Highline Community College
Hospitality and Tourism
Management
P.O. Box 98000
Des Moines, Washington
98198-9800

North Seattle Community College
Hospitality and Foodservice/
Culinary Arts
9600 College Way North
Seattle, Washington 98103

Olympic College
Foodservice Program
16th & Chester
Bremerton, Washington 98310

Pierce College
  Foodservice Management
  9401 Far West Drive S.W.
  Fort Lewis, Washington 98433

Renton Vocational Technical
    Institute
  Culinary Arts
  3000 N.E. Fourth Street
  Renton, Washington 98056

Seattle Central Community
    College
  Hospitality/Culinary Arts
  1701 Broadway, M/S 2120
  Seattle, Washington 98122

Skagit Valley College
  Foodservice Hospitality
  2405 College Way
  Mount Vernon, Washington
    98273

South Seattle Community College
  Culinary Arts
  6000 16th Avenue S.W.
  Seattle, Washington 98106

Spokane Community College
  Hotel, Motel, Restaurant
    Management
  N. 1810 Greene Street
  Spokane, Washington 99207

West Virginia

Community College of West
    Virginia State
  Hospitality Management
  Campus Box 183
  Institute, West Virginia 25112

Fairmont State College
  Foodservice Management
  Locust Avenue
  Fairmont, West Virginia 26554

Garnet Career Center
  Foodservice and Hospitality
  422 Dickinson Street
  Charleston, West Virginia
    25314

James Rumsey Vocational
    Technical Center
  Culinary Arts
  Route 6, Box 268
  Martinsburg, West Virginia
    25401

Shepherd College
  Hotel, Motel and Restaurant
    Management
  King Street
  Shepherdstown, West Virginia
    25401

Wisconsin

Chippewa Valley Technical
    College
  Restaurant and Hotel
    Cookery/Hospitality
    Management
  620 West Clairemont Avenue
  Eau Claire, Wisconsin 54701

Fox Valley Technical College
  Restaurant and Hotel
    Management and Cookery
  1825 North Bluemound Drive
  Appleton, Wisconsin 54913

Gateway Technical College
  Hotel, Motel Management
  1001 South Main Street
  Racine, Wisconsin 54303

Madison Area Technical College
  Culinary Trades
  3550 Anderson Street
  Madison, Wisconsin 53704

Mid-State Technical College
  Food and Hospitality
    Management
  500 32nd Street North
  Wisconsin Rapids, Wisconsin
    54494

Milwaukee Area Technical
    College
  Restaurant and Hotel Cookery
  700 West State Street
  Milwaukee, Wisconsin 53233

Moraine Park Technical College
  Restaurant and Hotel Cookery
  235 North National Avenue
  Fond du Lac, Wisconsin 54935

Nicolet Area Technical College
  Hospitality Management/Food
    Preparation
  P.O. Box 518
  Rhinelander, Wisconsin 54501

Southwest Wisconsin Technical
    College
  Foodservice Management
  Highway 18 East
  Fennimore, Wisconsin 53809

Waukesha County Technical
    College
  Hospitality Management/
    Culinary Arts
  800 Main Street
  Pewaukee, Wisconsin 53072

Western Wisconsin Technical
    College
  Foodservice Management
  6th and Vine
  La Crosse, Wisconsin 54601

Wisconsin Indianhead Technical
    College
  Hospitality Management
  2100 Beaser Avenue
  Ashland, Wisconsin 54806